Science Crossword Puzzles

Grades 6–12

Written by Rebecca Stark

ISBN 978-1-56644-565-8

Educational Books 'n' Bingo

Printed in the United States of America.

TABLE OF CONTENTS

*An alphabetical list of possible answers from which to choose is provided for each crossword puzzle. Use these lists at your discretion.

Earth Science

ACROSS

1 Large cave formed by the dissolution of soluble rocks (2 words)
3 Deep valley between cliffs, often carved from the landscape by a river
5 Vent in Earth's crust through which magma and steam erupt; structure resulting from these eruptions
7 Hot spring characterized by an intermittent turbulent discharge of water
8 Bends in a layer of rock caused by forces within the crust; usually occur in a series
11 Small openings from which hot gases and vapors are released into the air
16 Collective mass of water found on, under, and over the surface of a planet
17 Called magma within the mantle or crust; called lava when it reaches the surface (2 words)
18 Highly viscous layer directly under the crust and above the outer core
19 Earth's lithosphere is divided into 6 huge rigid ones and several smaller ones (2 words)
20 Result from the sudden breaking of rock within the earth (2 words)
21 Carrying away, or displacement, of solids, usually by wind, water, ice or waves
22 Describes Earth and other planets with a central metallic core and a silicate mantle
23 Sub-disciplines include mineralogy, paleontology and petrology
24 Large crater formed by a volcanic explosion or by the collapse of a volcanic cone
25 Continuous movement of water on, above and below the surface of the earth (2 words)
27 Underground bed or layer of Earth, gravel, or porous stone that yields water
28 Crust and uppermost mantle
29 Long, high sea wave caused by an earthquake, underwater landslide, or other disturbance
30 Lowest densest part of the earth's atmosphere in which most weather changes occur
33 Landforms that extend above the surrounding terrain in a limited area and have a peak
34 General term for material produced by volcanic eruption, especially ash
36 Rocks are aggregates of these

DOWN

2 Area of frequent earthquakes and volcanic eruptions; encircles basin of the Pacific Ocean (3 words)
4 Igneous, sedimentary and metamorphic are the three basic types
6 Disintegration of rocks into small soil particles due to rain, wind, temperature changes and other agents
9 A slowly moving mass of ice
10 Consolidated rock beneath the planet's surface
12 Mass of air surrounding the Earth, composed largely of oxygen and nitrogen
13 Movement of the continents relative to one another (2 words)
14 Fracture in the continuity of a rock formation caused by a shifting or dislodging of Earth's crust
15 Outermost solid shell of a planet or moon
24 Earth's is composed primarily of iron and some nickel
26 Caused by a sudden release of energy in the Earth's crust that creates seismic waves
31 Scale for expressing the magnitude of an earthquake
32 Large, fairly level highland areas separated from surrounding land by steep slopes
35 Rising and falling of Earth's oceans and other bodies of water caused mostly by gravitational forces of moon

Earth Science

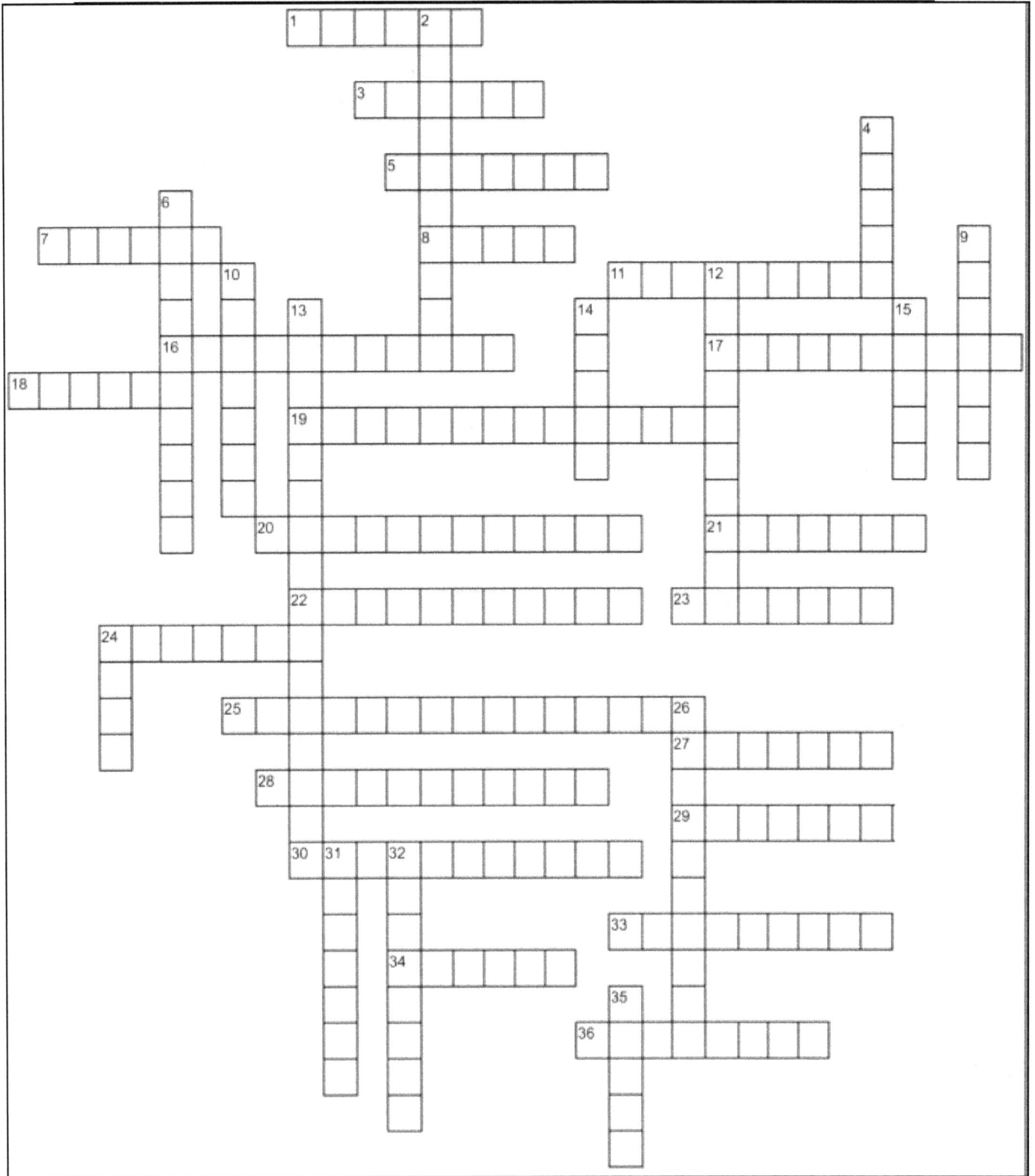

Science Crossword Puzzles: Grades 6 & Up

Life Science

ACROSS

6 Branch of science that studies animal life

7 Basic structural and functional unit of all living things

9 At risk of becoming extinct

11 Plant-like, photosynthetic organisms that grow in water; seaweeds are complex forms

12 Process of change by which species develop from preexisting species over time

13 Allows organism to survive and reproduce more effectively in its environment

17 Single-celled organisms with a cell membrane and cytoplasm but no cell nucleus

18 Branch of science that studies organisms too small to be seen with naked eye

19 Genetic material; acronym for "deoxyribonucleic acid"

20 Scientific name for humans (2 words)

21 Animal that lacks a vertebral column, or backbone

22 Biological process by which new individual organisms are produced

24 A living thing that influences or affects an another organism or the ecosystem (2 words)

28 Cannot be classified as a plant, animal, or fungus; mostly unicellular but algae are multicellular

29 Science concerned with the study of plant and animal life

30 Science of heredity

32 Molds, mildews, yeasts and mushrooms are examples

DOWN

1 Humans, apes and monkeys

2 Second part of binomial nomenclature, naming system established by Linnaeus

3 Branch of science that studies plant life

4 Feeding relationships between species within an ecosystem (2 words)

5 Single-celled animal-like organisms; more complex than bacteria

8 Animals with a backbone; fish, birds, amphibians, reptiles and mammals

10 Second stage in a butterfly's life cycle

14 Process that plants use to take energy from sun and convert it into a storable form

15 First part of binomial nomenclature, naming system established by Linnaeus

16 Sub-microscopic infectious agents; require a host cell to grow and reproduce

18 Warm-blooded vertebrate with hair

23 Unlike animals, they lack organs for mobility

25 A group of cells that have a similar structure and which function together as a unit

26 All living things are based on this element

27 Its aquatic larval stage is called a tadpole

30 Basic biological unit of heredity

31 A group of organs working together

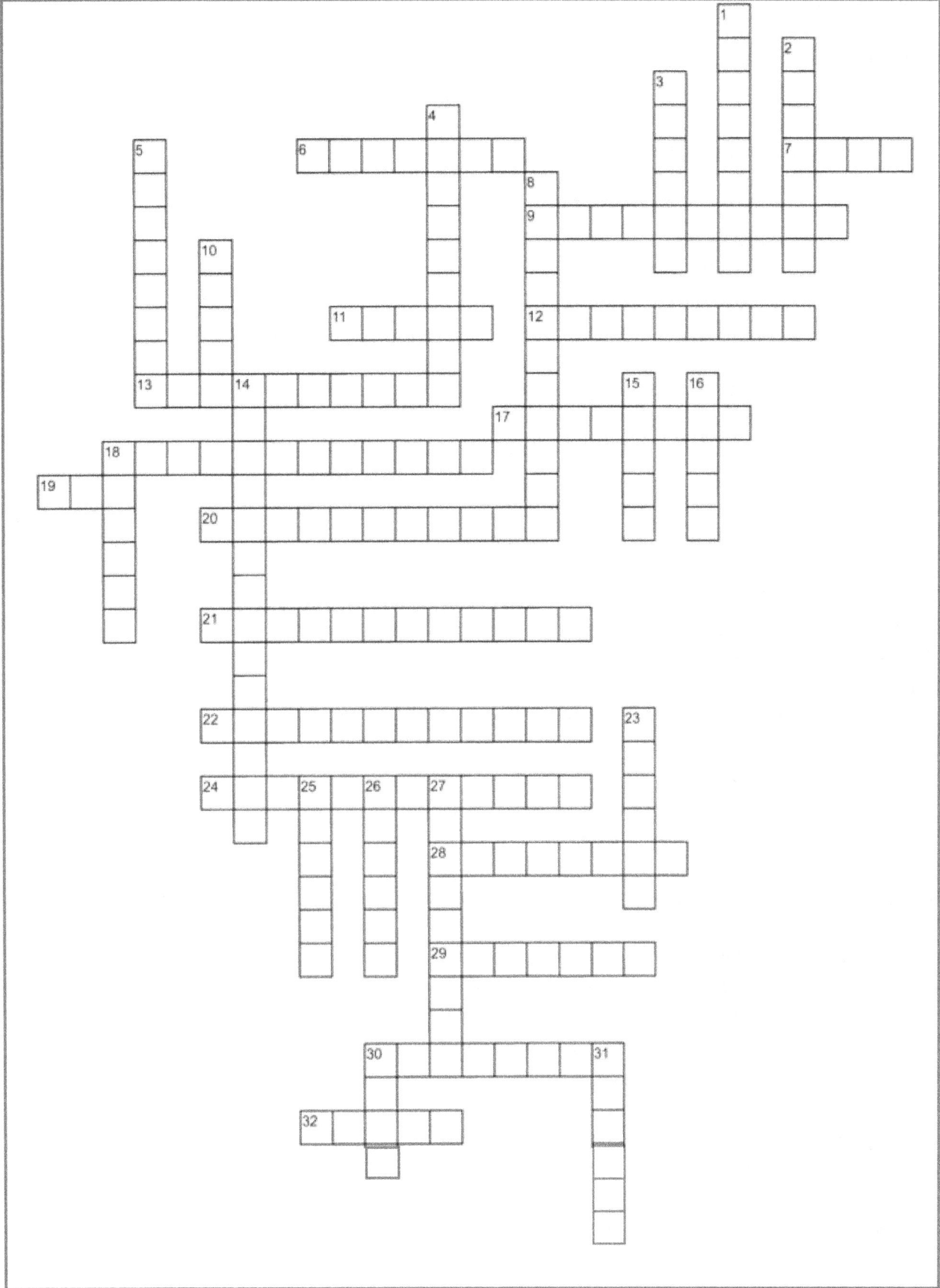

Life Science

Physical Science

ACROSS

4   Rate of doing work

6   A unit of power; equals one joule per second

7   Object used to disperse light and break it up into a spectrum

8   Material that produces a force which attracts iron, nickel or cobalt

14   Complete this analogy: fission : split :: ___ : join

16   Its 3 states are solid, liquid and gas

19   Wavelength of light determines how this is perceived

20   Measure of how closely packed together atoms of an element or molecules of a compound are

21   Flow of electrons in closed loops, or circuits; results from conversion of other energy source

22   Return of light, sound or other waves from a surface

23   Green plants convert this type of energy into chemical energy through photosynthesis

25   Science of matter and energy and their interactions

26   Bending of light as it passes from one substance to another

29   Degree or intensity of heat present in a substance or object

31   Transfer of heat energy in a gas or liquid by movement of currents

36   Science that deals with study of sound

38   Property that causes matter to have weight in gravitational field

39   Heat energy is energy in transit; ___ energy is internal energy in a system due to its temperature

DOWN

1   Periodic ones are characterized by crests, or highs, and troughs, or lows

2   Measurement of the amount of energy

3   Branch of physical science that studies the physical properties of light

5   The number of vibration per second is the ___ of the sound

9   Compound machine that converts mechanical energy into electrical energy

10   Describes stored energy

11   Material that allows electricity or heat to flow through it easily

12   Simplest structural unit of an element or compound; made up of atoms

13   Object's ability to float

15   Branch of science that deals with composition, properties and reactions of substances

17   Toothed wheel

18   Study of the motion of air as it interacts with a moving object

24   Push or pull that causes a change in the motion of an object

27   Smallest component of an element

28   Force that stops stationary things from moving and slows down things already in motion

30   Energy an object possesses due to the vibration of its molecules

32   Type of energy resulting from fission or fusion of the nuclei of atoms

33   Power derived from the utilization of physical or chemical resources

34   Material that resists flow of electric current or heat

35   The part of an atom that contains protons and neutrons; electrons move around it

37   Distribution of colors produced when white light is separated into its constituent colors

Physical Science

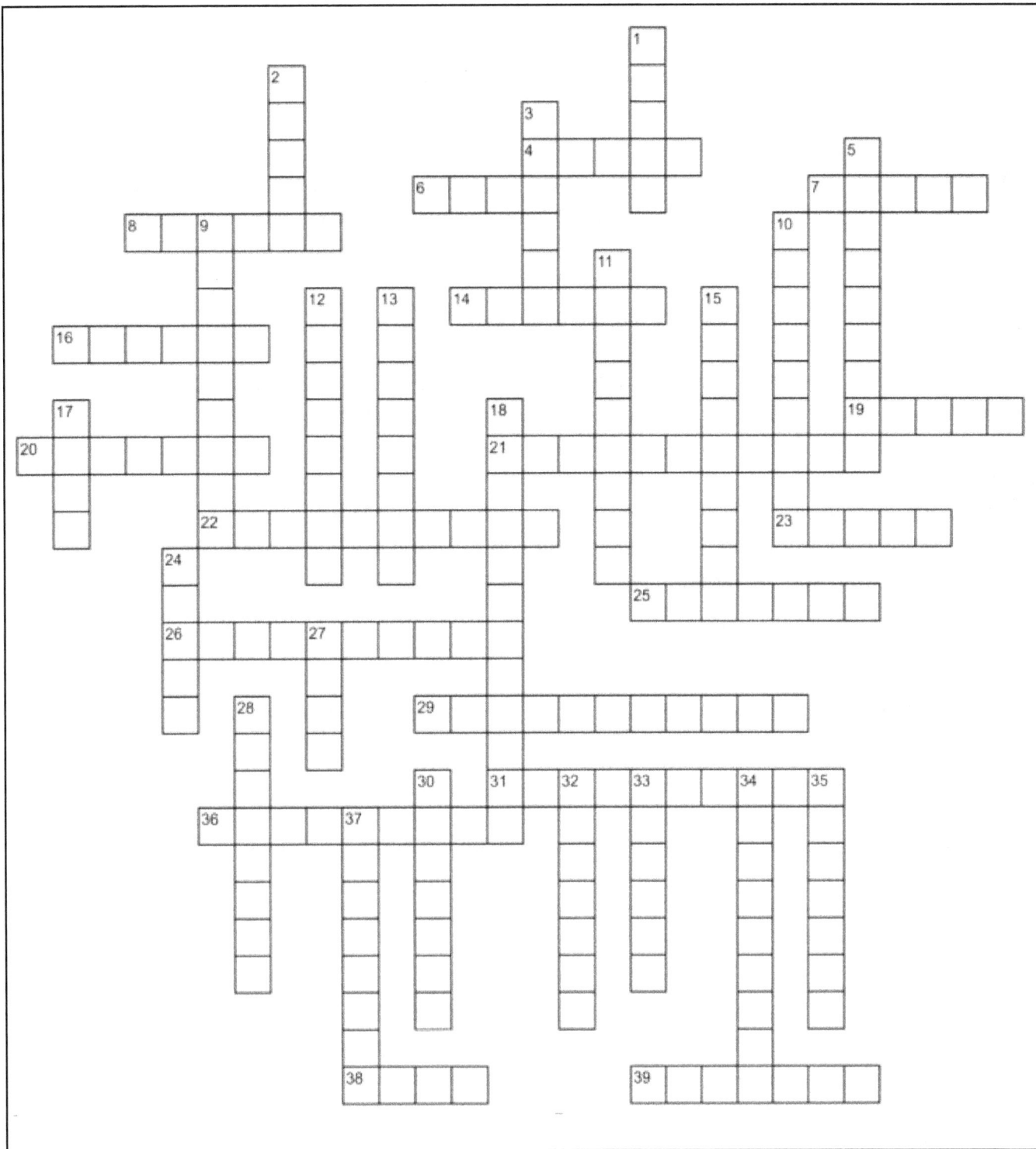

Science Crossword Puzzles: Grades 6 & Up

Astronomy

ACROSS

1   Its status as the ninth planet keeps changing
5   Acronym for National Aeronautics and Space Administration
6   Thought to be a rapidly rotating neutron star; emits regular pulses of radio waves
8   Second planet from sun; named for the ancient Roman goddess of love
10 Measures angular distances between objects; useful for taking altitudes in navigation
11 Collection of gas, dust, and billions of stars and their solar systems; held together by gravity
12 Class of giant planets including Jupiter, Saturn, Neptune and Uranus
13 Closest planet to sun and smallest planet in our solar system
15 Large meteor that explodes in the atmosphere
16 One who studies objects outside of Earth's atmosphere
17 Ours comprises 8 planets* and their moons as well as smaller bodies (2 words)
18 Small bodies that travel through space; most are pieces broken off from an asteroids
21 Results from the collapse of a star (2 words)
24 Third planet from sun
25 Class of rocky planets including Mercury, Venus, Earth, and Mars
27 Small metallic and rocky bodies that orbit the sun between the orbits of Mars and Jupiter
28 Sixth planet from sun and second largest; has spectacular system of ringlets
29 A planet's natural satellite
32 Eighth planet from sun; named after Roman god of sea
33 Large, round body that orbits the sun
35 ___ effect is noticeable change in the frequency of sound, light or water waves as the source and the observer move

DOWN

2   Like Neptune, this seventh planet from the sun is an ice giant
3   A massive and extremely remote celestial object, emitting exceptionally large amounts of energy
4   Parts include a nucleus, a coma, a hydrogen envelope, a dust tail, and an ion tail
7   Comprises all the space and matter in existence
9   Occurs when one object in space blocks another from view
10 Any object that orbits a celestial body; moons are natural ones
14 Group of stars that seem to form imaginary outlines or meaningful pattern
19 Theory that a cosmic explosion marked the origin of the universe (2 words)
20 Largest planet in our solar system
22 Optical instrument designed to make distant objects appear nearer
23 Name of our galaxy (2 words)
26 Star that suddenly increases greatly in brightness due to a catastrophic explosion
30 Fourth planet from sun; sometimes called Red Planet
31 Our sun is one
34 Imaginary belt in the sky that includes the paths of the planets; divided into 12 constellations

* Controversy over whether or not Pluto should be considered a planet remains.

Astronomy

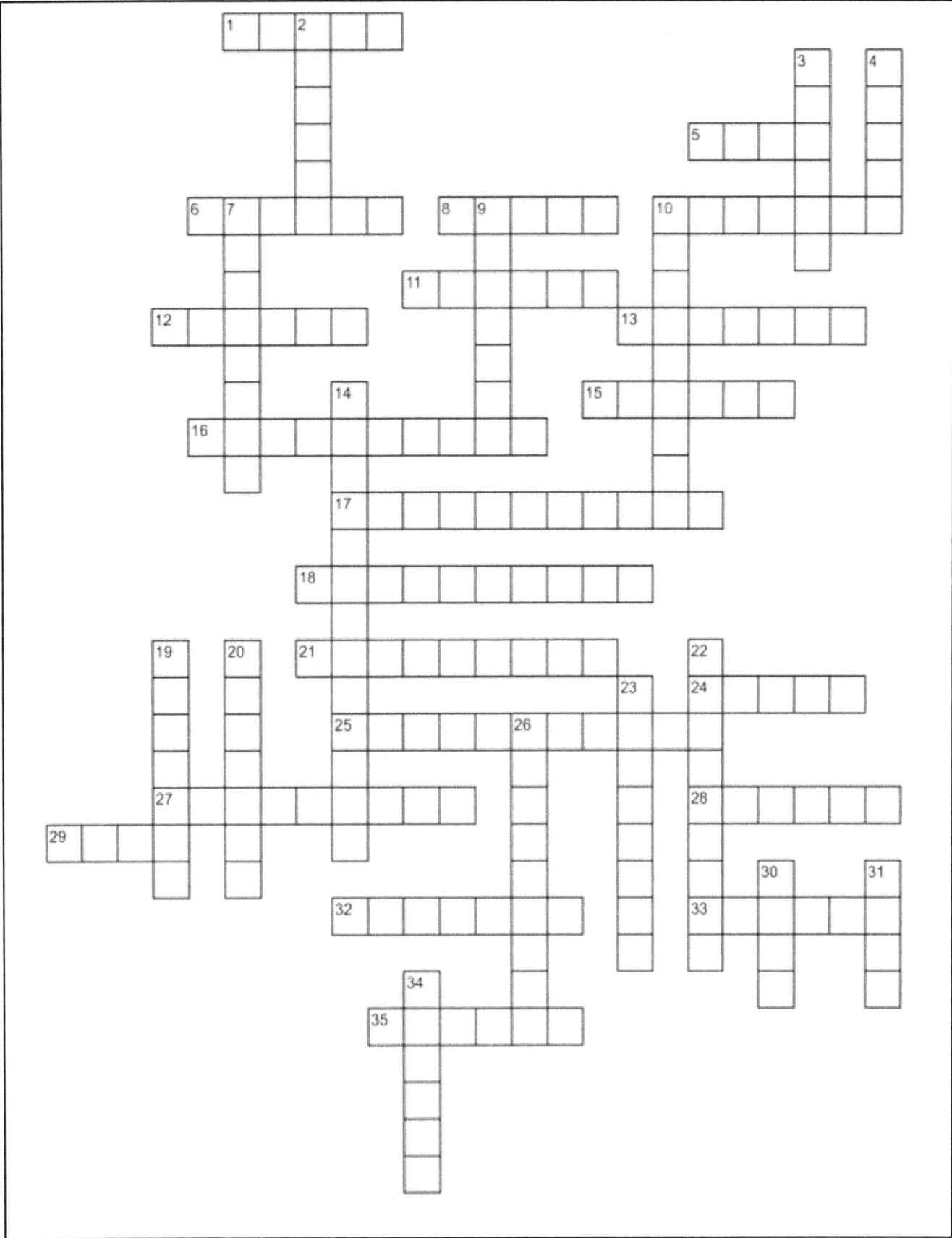

11

Ecology

ACROSS

5   Bring energy into ecosystem from inorganic sources; primary producers
8   Relationship between 2 organisms: 1 benefits and other is neither harmed nor helped
9   At first trophic level of food chain: _____ producers
11 Capable of being broken down by the action of living things
12 Largest ecosystems in the world
13 Process by which an organism becomes better suited to its environment
15 Organism that breaks down dead plants and animals
17 Variation of life forms in a specific region, such as an ecosystem
20 Organism that depends on host to survive; it benefits, but host is harmed
21 Forest with cone-bearing, needle-leaved or scale-leaved evergreen trees
22 Place where freshwater river or stream flows into the ocean, mixing with the seawater
23 Large community of plants and animals that occupies a distinct region
24 Permanent destruction of indigenous forests and woodlands
27 Lakes and ponds are examples of ___ ecosystems
30 Position an organism occupies in a food chain is its ___ level
31 A grassy plain in tropical and subtropical regions
32 Ecological role of an organism in its community
33 Terrestrial ecosystem also known as boreal forest or snow forest
34 Pertains to the living components of an ecosystem
35 The coldest biome

DOWN

1   Organism that feeds on refuse or carrion
2   Close ecological relationship between 2 or more individuals of different species
3   Describes trees that lose leaves in fall and regrow them in spring
4   Type of environment in which an organism or an ecological community usually lives
6   Ecosystem with tall trees, a warm climate, and at least 80 inches of rain per year (3 words)
7   A secondary consumer; a predator
10 Arid area of sparse vegetation
14 Found in warm, shallow marine waters; formed by the stony skeletons of living organisms (2 words)
16 Biotic and abiotic factors of an environment working as a unit
18 A primary one gets its energy by eating producers, or autotrophs
19 Effect caused by rise in temperature because gases in atmosphere trap energy
25 Gradual supplanting of one community of plants by another
26 Pertains to the physical and chemical components of the environment
27 Path of energy from one organism to the next within an ecosystem (2 words)
28 Swamps, marshes and bogs, for example
29 Study of interrelationships between organisms and their environment

Ecology

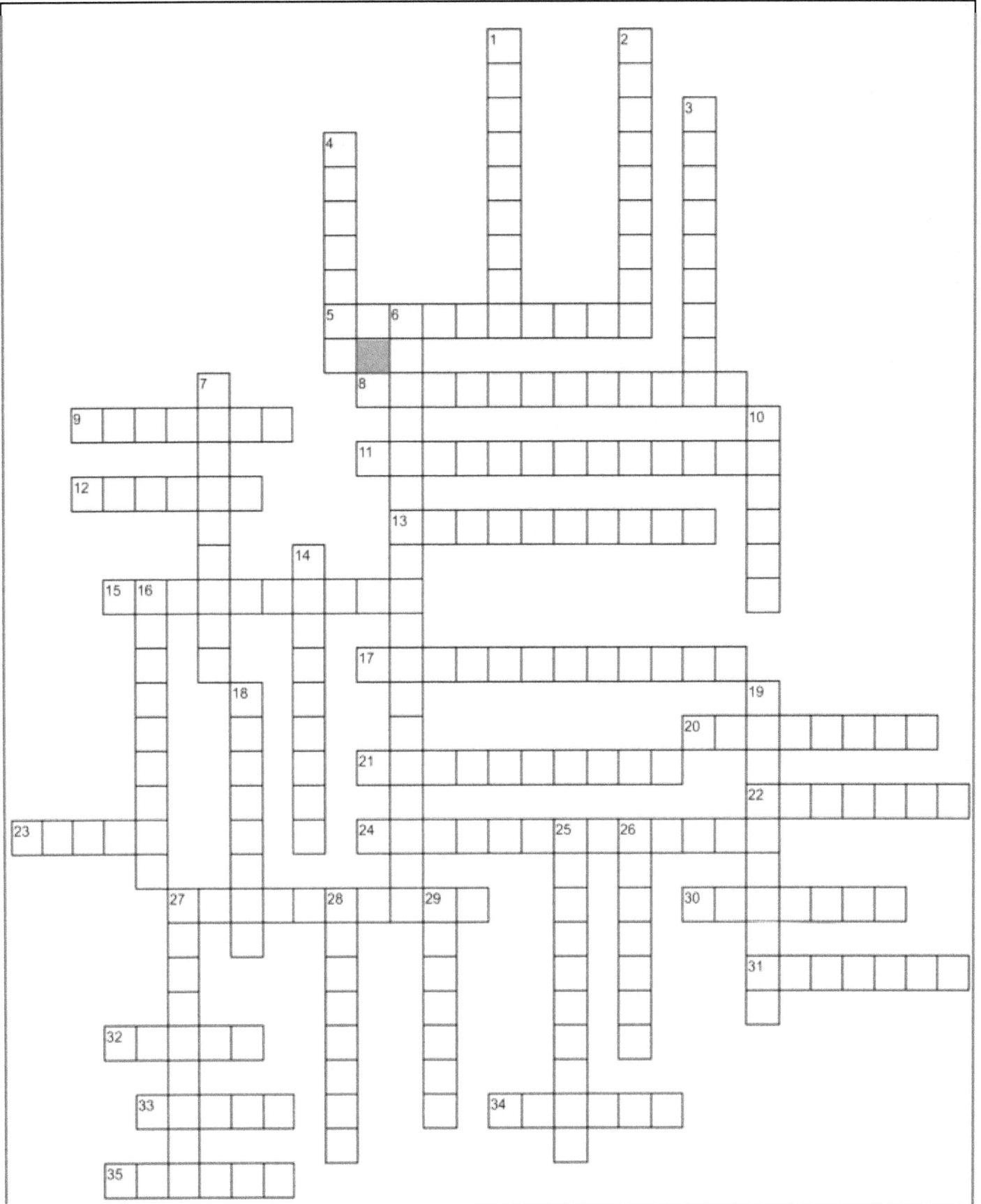

Science Crossword Puzzles: Grades 6 & Up

Oceanography

ACROSS

2   Large floating mass of ice detached from a glacier
6   Alternate rising and falling of the surface of the ocean
8   A ring of coral reefs that encircle a lagoon
11 At first level of the marine food chain
13 Giant, slow moving wave caused by the displacement of water due to underwater earthquake or volcano
15 Acronym for self-contained underwater breathing apparatus
16 Body of salt water that covers more than 70 percent of Earth's surface
17 Top zone of the ocean; also called euphotic zone
18 Plant-like, mainly aquatic, organisms; seaweeds and kelp are examples
19 Plants and animals that live in, on, or near the bottom of a body of water
20 Rigid plates that move slowly but continuously over the upper part of the mantle (2 words)
23 Describing fish like sharks, skates and rays
25 Pacific Basin called this because about 75% world's volcanoes occur in its rim (3 words)
27 Describing region of lake or ocean where there is little or no sunlight
30 Scientific study of the ocean and its phenomena
33 Relative concentration of salt
34 Animals that swim freely in the ocean
35 Ocean once considered southern region of 3 other oceans
36 Zone in an ocean or lake that receives sunlight; its uppermost part where photosynthesis can occur is called the euphotic zone
37 Largest, deepest and oldest of the oceans

DOWN

1   One of the series of ridges that move across the surface of the ocean
2   Zone above water at low tide and under water at high tide
3   Ability of something to float
4   A clam is one; so is a snail
5   Water that remains on a shore or reef after the tide recedes (2 words)
7   Microscopic marine animals; krill is an example
9   Light produced by chemical reaction within an organism
10 Large, carnivorous, aquatic mammals; whales and dolphins are examples
12 Steady flow of the ocean's surface water in a prevailing direction
14 Porous primitive life forms; water flows in and out of body
21 Of or relating to the sea
22 Reef formed from tiny animals, called polyps, that live in colonies
24 In the North Atlantic; so named because of the seaweed that floats on its surface (2 words)
26 Located between Africa, the Southern Ocean, Asia and Australia
28 Comprise more than 29,000 species; salmon and tuna are examples (2 words)
29 Smallest and most shallow ocean
31 Second-largest ocean; borders mainland of USA to the east; relatively shallow
32 Microscopic plants and animals that drift in the water

Oceanography

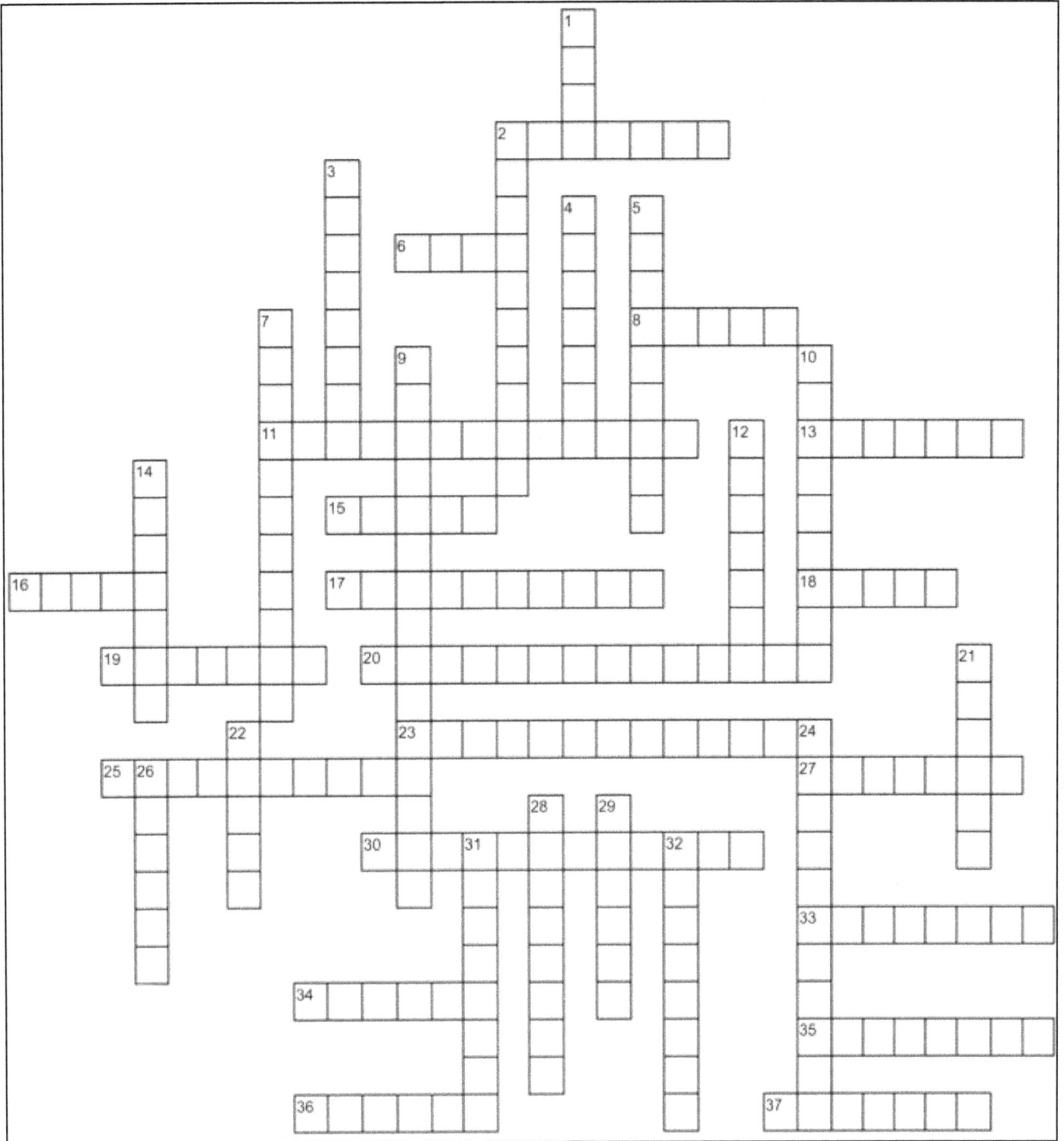

Psychology

ACROSS

1   Type of mechanisms that help people cope with reality; repression and projection are two
7   Known for his theory of multiple intelligences
8   Learned negative attitude toward objects or people
9   An irrational and persistent fear of a situation, activity, thing, or person
10 Study of mind and mental processes, especially in relation to behavior
11 The actions and reactions of an organism as it adjusts to its environment
13 Process by which we incorporate new information into our existing ways of thinking
15 Depression and bipolar disorder are examples of this type of disorder
16 Primary ones include anger, fear, grief and joy
19 Unwanted entrance into it by another person can make an individual feel uncomfortable (2 words)
23 Occurs when one is blocked from reaching a personal goal
24 Process of changing existing ideas to adapt to new information
25 Ability to learn from experience, adapt to new situations, understand abstract concepts, and use knowledge
26 A mental condition, such as schizophrenia, that causes a person to lose touch with reality
27 Thorndike, Father of Educational Psychology, said animals learn this way (3 words)
28 Insomnia is this type of disorder
29 Person with an antisocial personality disorder: a pathological liar, lacks remorse, and can be charming if for self-gain
30 Skinner demonstrated operant ___ by rewarding a pigeon with a pellet of food
34 Disorders characterized by unwarranted worry and tension
35 Believed human personality comprised the id, the ego and the superego
36 Lasting change in behavior that is the result of experience; conditioning
37 Capacity to understand another's state of mind or emotion

DOWN

2   Personality disorder characterized by grandiosity, lack of empathy, and need for excessive admiration
3   Degree to which members of a group will alter their behavior and views to fit those of the group
4   Hostile behavior intended to cause harm
5   Process of expressing strong, but repressed emotions
6   Redirection of unconsciously retained feelings toward a new object
12 Processes that include perception, language, problem solving and abstract thinking
14 Individual's acceptance of established norms
17 Example of an Altered State of Consciousness
18 Mnemonic device that organizes single items into units
20 Coined term "collective unconscious"
21 Divided into 3 storage systems: sensory, short term and long term; amnesia is lack of it
22 Hypochondriacs are abnormally anxious about their ___
23 Functional ___: tendency to see objects as only working in a particular way
26 Combination of behavior, emotion, motivation, and thought patterns that define an individual
31 False beliefs held in spite of evidence to the contrary; persecutory ones are most common
32 Unlike hallucinations, these are misinterpretations of real sensations
33 Self-actualization is final level of psychological development In Maslow's hierarchy of these

Psychology

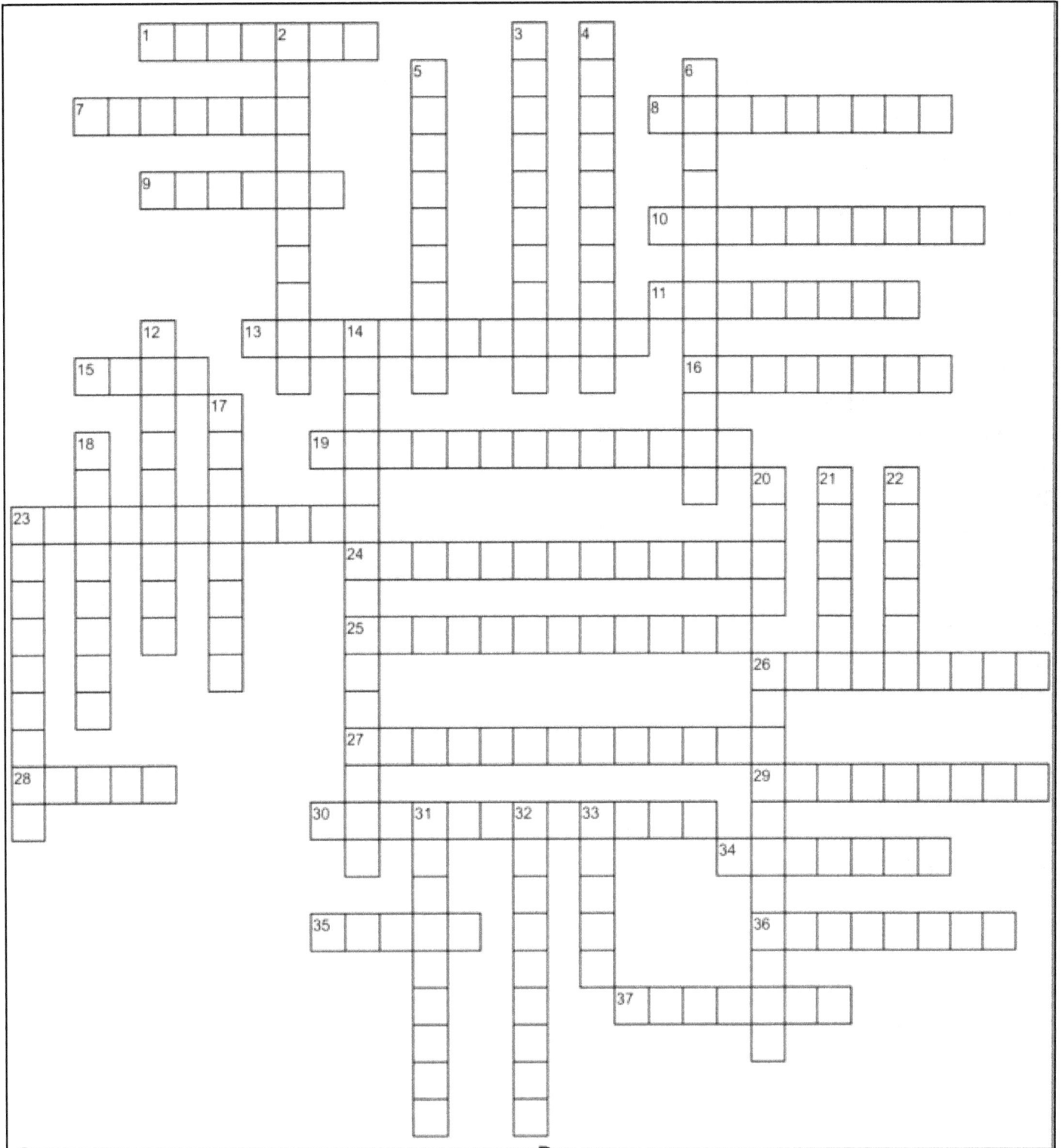

Genetics and Heredity

ACROSS

4   Basic unit of heredity
6   Allele that does not produce characteristic effect when dominant allele is present
7   Change in DNA sequence resulting in creation of new trait not in parental type
10 Set of genes in our DNA which is responsible for a particular trait
11 To receive a characteristic from one's parent by genetic transmission
12 ABO is classification system for this (2 words)
15 DNA and RNA (2 words)
18 Having two different alleles for the same trait
21 Form of artificial DNA created by combining 2 or more sequences
23 Mature sexual reproductive cell
25 When neither allele of a gene pair is dominant or recessive
26 Chromosomes are found in their nucleus
28 Building blocks of nucleic acids: uracil, cytosine, thymine, adenine and guanine
30 Passing on of traits from parents to their offspring; its study is called genetics
31 Diagram showing genealogy of an individual and direct ancestors to follow inheritance of trait
33 Allele or gene that masks the effect of the recessive allele or gene when present
34 Cell division that reduces the number of chromosomes to half
35 Cell division in which nucleus divides into 2 nuclei with same number of chromosomes as parent
36 One of a pair or series of genes that occupy specific position on specific chromosome
37 Hereditary material in most organisms; deoxyribonucleic acid

DOWN

1   Part of cell where DNA is found
2   A structure in all living cells that carries the genes determining heredity; most cells in the human body have 23 pairs
3   Hereditary blood-coagulation disorder
5   Scientist who studies genetics
8   Complete set of genetic material in an organism
9   Called Father of Genetics (last name)
13 Diagram used to predict outcome of a particular cross or breeding experiment (2 words)
14 A cell or organism that has a pair of each type of chromosome
16 Characteristics of an organism
17 Entire physical, biochemical and physiological makeup of an individual
19 Single cell that contains the genetic material of both the mother and father
20 Branch of biology that studies heredity
22 Organism genetically identical to the individual from which it was derived
24 Having two identical alleles for the same trait
27 Adjective describing something genetically transmitted from parent to offspring
29 Describes monozygotic twins
32 Ribonucleic acid; transfers genetic code from DNA to protein-forming system of the cell

Genetics and Heredity

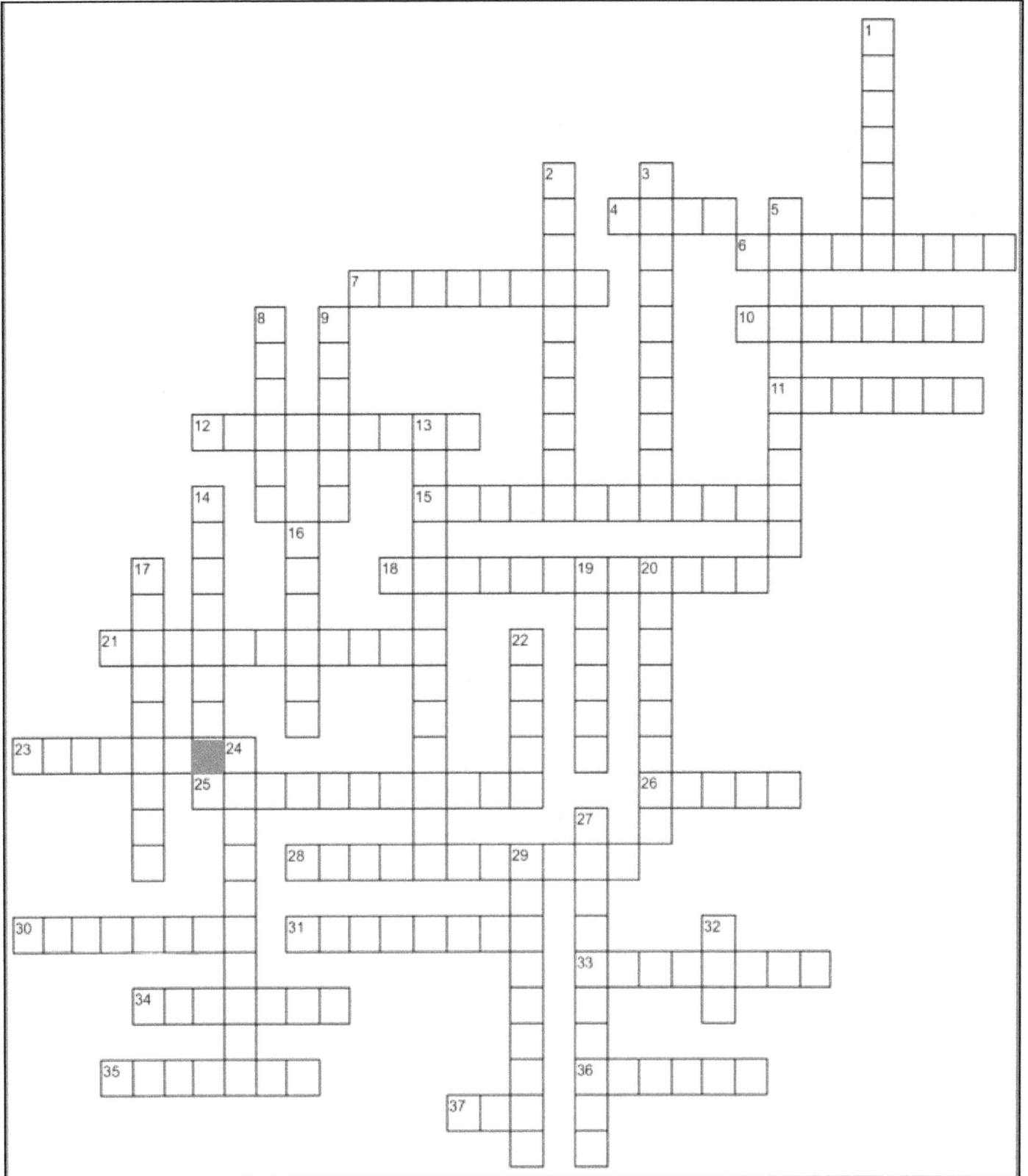

Crime Scene Investigation

ACROSS

2   Acronym for Automated Fingerprint Identification System

7   Shoe prints and tire tracks are examples of this type of evidence

11   An old, unsolved case is said to be this

12   Internal and external examination of a body after death; postmortem

14   Application of science to the criminal and civil laws enforced in criminal justice system

16   Unburned primer powder sprayed on hands of someone firing a gun (2 words)

17   Path a moving object follows through space

18   In a murder case, it is the body of the victim; means "body of crime" (2 words: Latin)

20   A list that records every official person who handles a piece of evidence (3 words)

21   Individual who may have committed the crime

22   Pattern of blood that has struck a surface (2 words)

24   Cord-like object used for strangulation

26   Act of ruining evidence by accidentally depositing outside trace evidence

27   Forensic science of analyzing and interpreting evidence

29   A method of separating and analyzing mixtures of chemicals.

32   Acronym for system used to share FBI's DNA profiles with law enforcement bodies

33   Forms chains of genetic material organized into chromosomes

34   Anything used, left, removed, altered or contaminated during commission of the crime

35   Investigates suspicious deaths; may or may not be forensic pathologist; appointed or elected

36   Compound that reacts to red blood cells and gives off a blue-greenish light

37   Present but not visible

38   Something false used to deceive; signing someone's name to document without permission

39   Forensic ___ is the study of insects and arthropods in relation to a criminal investigation

DOWN

1   Stiffening of the body that begins about 30 minutes after death (2 words: Latin)

3   Basic ___ patterns include loops, ridges and whorls

4   Coloration of skin of lower parts of corpse caused by settling of red blood cells (2 words: Latin)

5   Acronym for Forensic Information System for Handwriting

6   Adjective meaning "before death"

8   Usual method of operation used by a perpetrator (2 words: Latin)

9   Postmortem cooling of a dead body (2 words: Latin)

10   Study of motion of bullets and distinctive characteristics after being fired

13   Internal diameter of the gun barrel or of the projectile it shoots

15   Unlike a coroner, this professional is always a trained medical practitioner (2 words)

19   Individual who committed the crime

23   Locard's ___ principle states that "every contact leaves a trace'

25   Describes material such as hairs and fibers that are accidentally deposited at a crime scene

28   Essential for establishing line of security around crime scene (2 words)

29   Science that explores the nature and prevention of crime

30   Legal classification of how someone died: suicide, natural, accidental or homicide (3 words)

31   Forensic ___ is application of dentistry to investigation of crime

Crime Scene Investigation

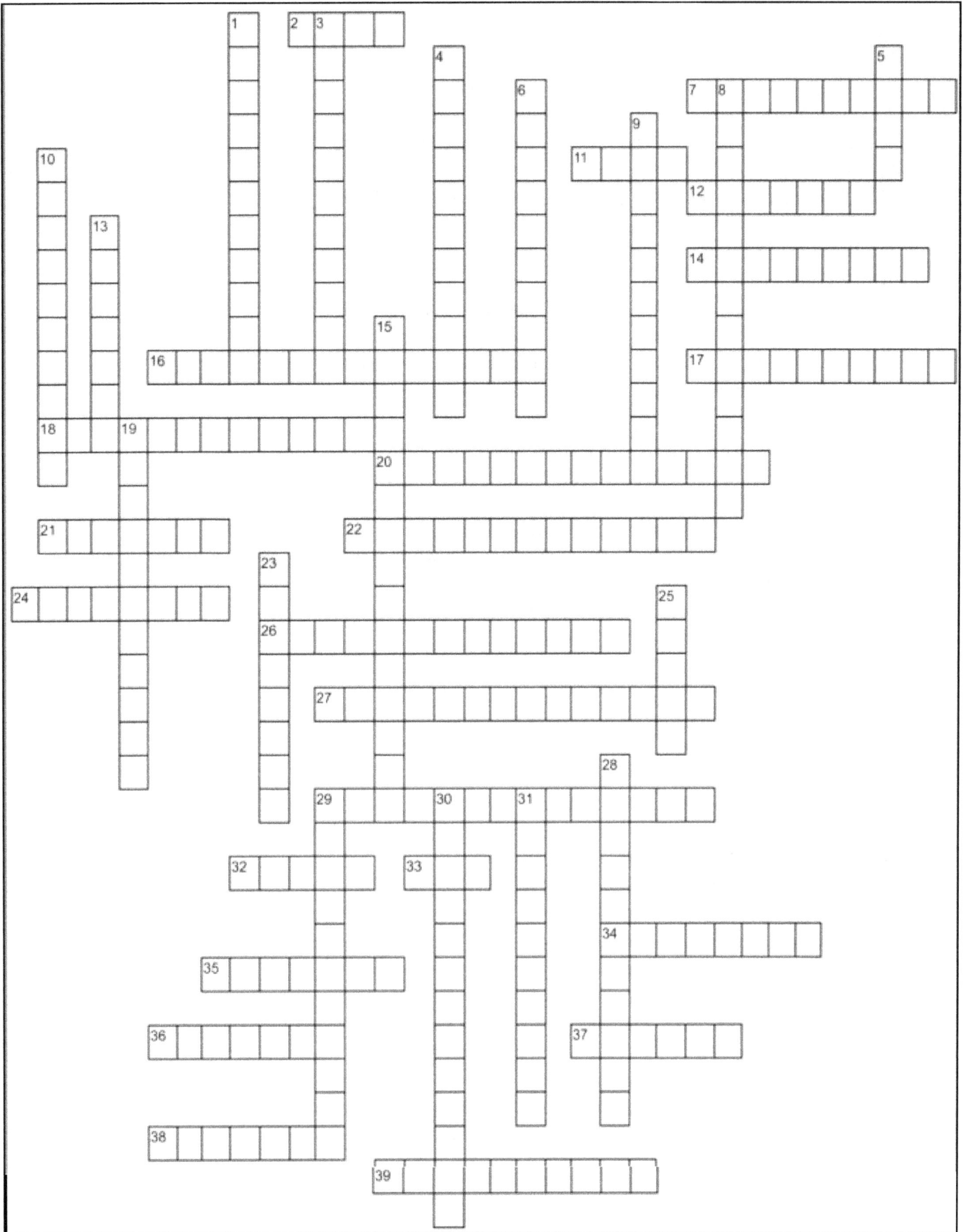

The Human Body

ACROSS

2 Brain and spinal cord make up central ____ system
5 Breathing organs that remove carbon dioxide from blood and brings oxygen to it
8 Structure at which two parts of the skeleton are fitted together
10 Group of tissues that performs one or more specific functions
11 Basic biological unit of heredity
12 Chemical substances that regulate activity of certain cells or organs
13 Organ that pumps blood throughout the body
14 Science of heredity
16 Smallest blood vessels
19 System comprising heart, blood and blood vessels; also called vascular or cardiovascular.
21 System that includes lungs, trachea, bronchi and diaphragm
23 Blood vessels that carry deoxygenated blood back to the heart; not as strong as arteries
24 System responsible for breaking down food into nutrients
25 Genetic material; acronym for "deoxyribonucleic acid"
28 Biological process by which new individual organisms are produced
29 System that removes excess materials from body fluid; includes urinary system
31 Another name for the circulatory system
34 System containing glands that help the body maintain a stable environment
35 A group of cells that have a similar structure and which function together as a unit
36 Complete this analogy: cardiac : heart :: ____ : lungs
38 System that helps keep out harmful bacteria and viruses
39 A group of organs and related tissues working together

DOWN

1 Largest organ of the human body; acts as barrier against infection and provides a sense of touch
3 Its muscular walls push food down into the stomach
4 Adult human body has 206
6 System that provides support, allows movement and protects our vital organs
7 One of the bony segments of the spinal column
9 Connect branches of trachea to the lungs
15 Basic structural and functional unit of all living things
17 Comprises main organs of the digestive system: esophagus, stomach, and intestines (2 words)
18 Fluid circulating through heart, arteries, veins, and capillaries; carries oxygen to tissues
20 The only voluntary ones are the ones attached to our bone
22 Organ that controls the human nervous system
26 Filters the blood from the digestive tract before passing it to the rest of the body
27 One of the upper chambers of heart that receive blood from veins and push it into ventricles
30 An involuntary reaction
32 Divided into 2 hemispheres, joined at bottom by corpus callosum
33 Large blood vessels that carry blood away from the heart
37 Bundle of fibers that transmit information from one body part to another

The Human Body

Ecosystems Word Search

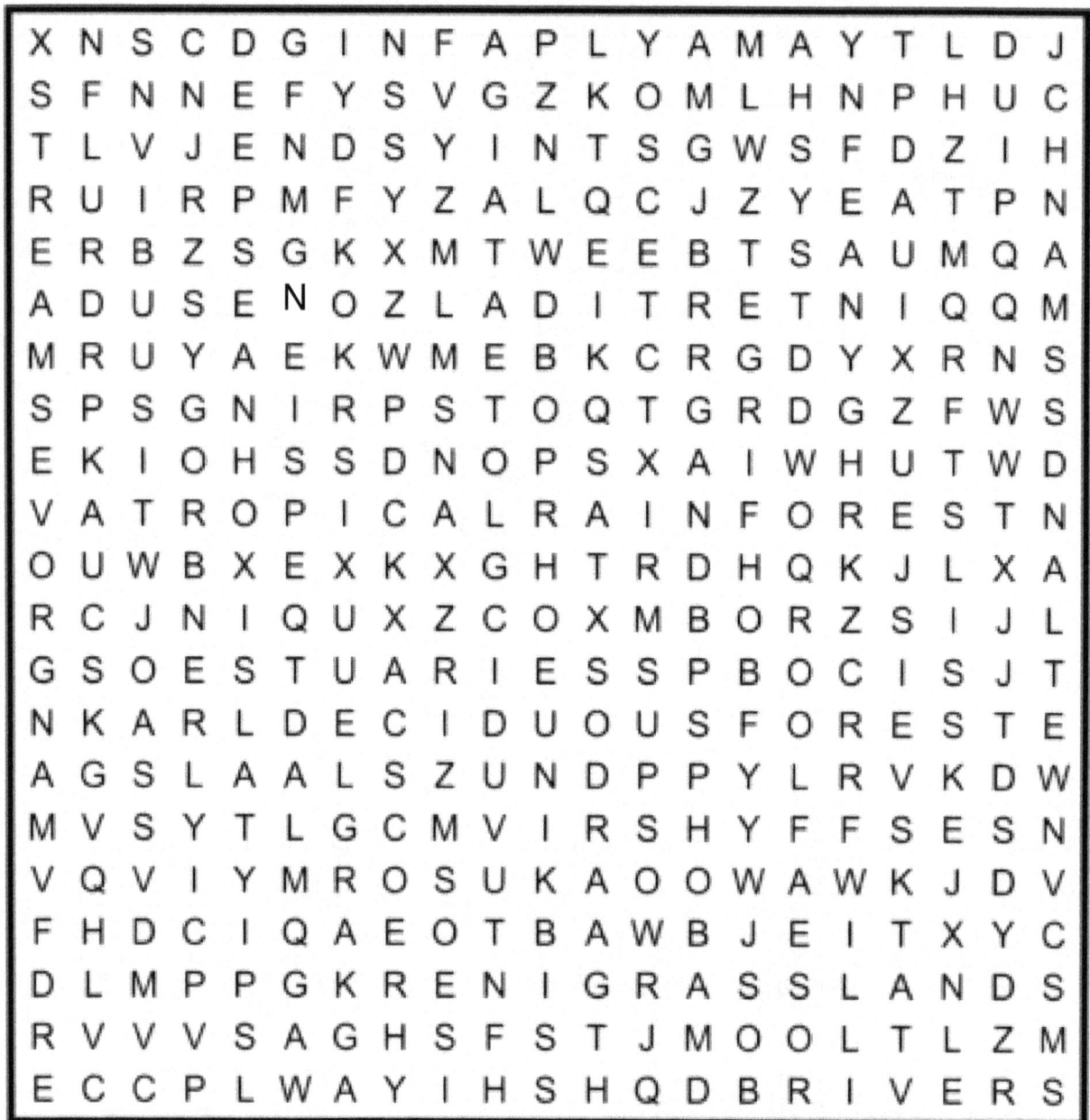

```
X N S C D G I N F A P L Y A M A Y T L D J
S F N N E F Y S V G Z K O M L H N P H U C
T L V J E N D S Y I N T S G W S F D Z I H
R U I R P M F Y Z A L Q C J Z Y E A T P N
E R B Z S G K X M T W E E B T S A U M Q A
A D U S E N O Z L A D I T R E T N I Q Q M
M R U Y A E K W M E B K C R G D Y X R N S
S P S G N I R P S T O Q T G R D G Z F W S
E K I O H S S D N O P S X A I W H U T W D
V A T R O P I C A L R A I N F O R E S T N
O U W B X E K X G H T R D H Q K J L X A
R C J N I Q U X Z C O X M B O R Z S I J L
G S O E S T U A R I E S S P B O C I S J T
N K A R L D E C I D U O U S F O R E S T E
A G S L A A L S Z U N D P P Y L R V K D W
M V S Y T L G C M V I R S H Y F F S E S N
V Q V I Y M R O S U K A O O W A W K J D V
F H D C I Q A E O T B A W B J E I T X Y C
D L M P P G K R E N I G R A S S L A N D S
R V V V S A G H S F S T J M O O L T L Z M
E C C P L W A Y I H S H Q D B R I V E R S
```

An ecosystem comprises the living (biotic) and nonliving (abiotic) components of a natural community. For the word search, find the words in each list.

Freshwater Ecosystems
LAKES
PONDS
RIVERS
STREAMS
SPRINGS
WETLANDS

Terrestrial Ecosystems
TUNDRA
TAIGA
DECIDUOUS FOREST
TROPICAL RAIN FOREST
GRASSLANDS
DESERTS

Marine Ecosystems
SALT MARSH
INTERTIDAL ZONES
ESTUARIES
LAGOONS
MANGROVES
CORAL REEFS
DEEP SEA
SEA FLOOR

Science Crossword Puzzles: Grades 6 & Up

Hidden Body Parts

Find the hidden body part in each sentence. A clue is provided for each.

1. Noel bowed to the king.
Clue: Joint between the forearm and the upper arm

2. Jack needs a bandage for his wound.
Clue: Joint between the thigh and the lower leg.

3. Hank led the team in home runs.
Clue: Joint connecting the foot with the leg

4. Sam asked his mother, "Should Eric or I go to the store?"
Clue: Upper joint of the human arm

5. Because of the strong wind, Piper did not open her umbrella.
Clue: What the trachea is sometimes called

6. The child was fascinated by the colony of ants.
Clue: What the large intestine is called

7. The livery car arrived late, so the couple almost missed their flight.
Clue: Organ that secretes bile and stores fat and sugar

8. Emma watched intently as two bears frolicked in her backyard.
Clue: Organs that perceive sound by detecting vibrations

9. Sandy does not like sweet or sour food.
Clue: Body without head, neck, and limbs

10. Sir isaac Newton is known for his laws of motion and universal gravitation.
Clue: Colored, ring-shaped membrane behind the cornea of the eye

11. When told to write about an adventure, Tina wrote about rock climbing.
Clue: Layer at the back of the eyeball containing cells that are sensitive to light

12. "Chris, kindly shut the door on your way out," the shopkeeper requested.
Clue: Organ forming the natural outer covering of the body

Volcanic Eruptions Puzzle

Complete the puzzle.
Then write the letters written on the larger emboldened lines on the lines at the left.
The result will be a word describing "rocks formed from the solid fragments
ejected during a volcanic eruption."

_____ A very light and porous volcanic rock   —— — — —— —

_____ A type of hot spring   — — —— — — —

_____ A large cave   — — — — —— —

_____ A lump of lava thrown out by a volcano   — —— — —

_____ Earth's is mostly iron   —— — — —

_____ Rock fragments ejected from volcano   —— — — —— — —

_____ Powdery material thrown out by a volcano   —— — —

_____ Hard, dark, glasslike volcanic rock   — — —— — — —— —

_____ A volcano is one in the Earth's surface   — — — ——

_____ Rocks formed by cooling & solidifying of magma or lava   —— — — —— — —

_____ Combustible rock; consists mainly of carbonized plant matte   —— — — —

Solutions*

*Optional Lists of Answers

Alphabetical lists of the answers are provided. These may be used to help solve the puzzle from the beginning, to assist those having difficulty, or not at all.

Earth Science

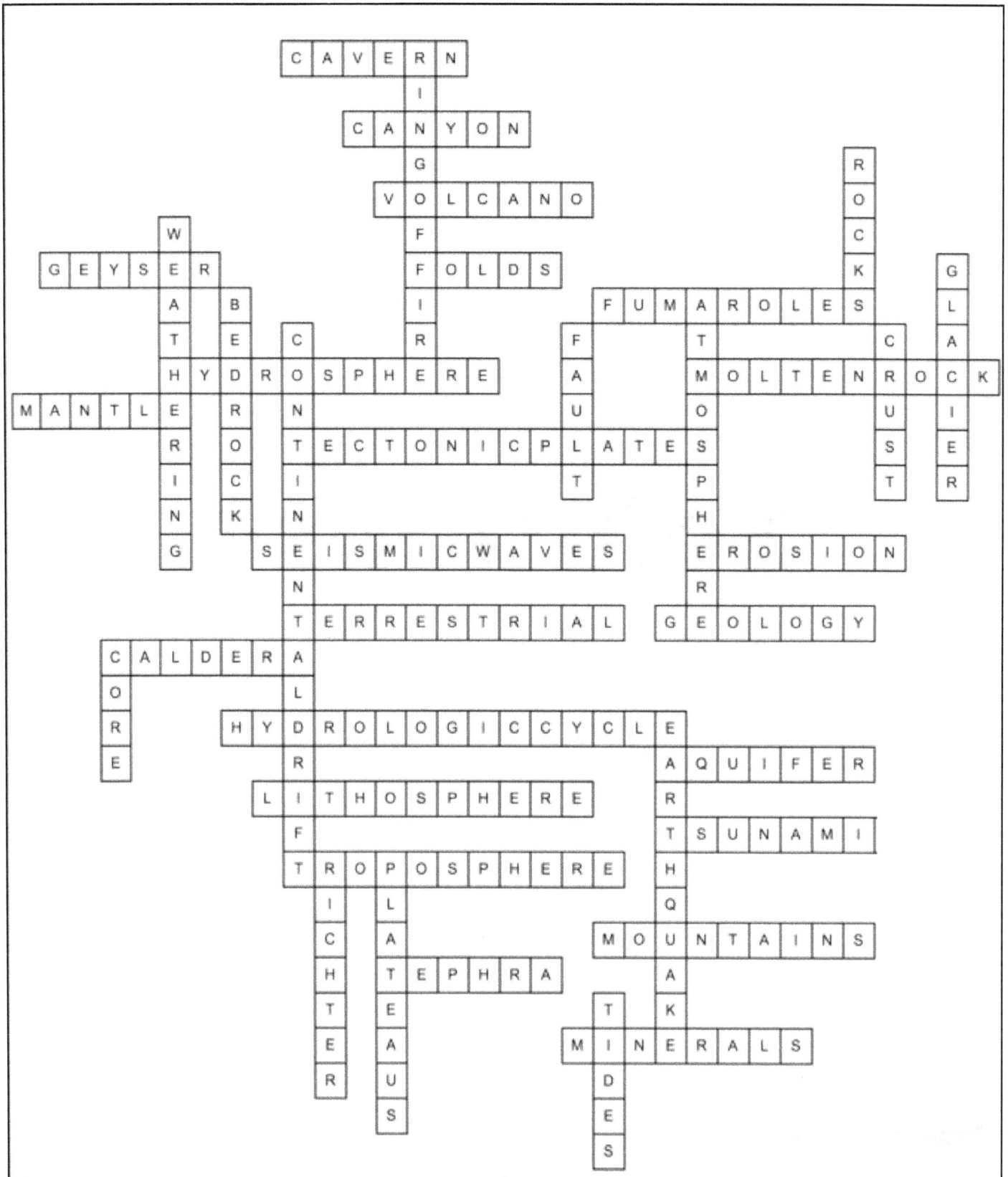

A crossword puzzle grid with the following answers filled in:

CAVERN

CANYON

VOLCANO

FOLDS

GEYSER

FUMAROLES

HYDROSPHERE

MOLTEN ROCK

MANTLE

TECTONIC PLATES

SEISMIC WAVES

EROSION

TERRESTRIAL

GEOLOGY

CALDERA

HYDROLOGIC CYCLE

AQUIFER

LITHOSPHERE

TSUNAMI

TROPOSPHERE

MOUNTAINS

TEPHRA

MINERALS

Down/intersecting words include: RINGING, OFFIRE, WATHRING, BE D ROCK, CONINTAT, SENTATER, CORE, DRIFT, LITHTER, ICHTER, ROCK, CRUST, GLACIER, FAULT, ATMOSPHERE, PLATEAUS, TIDES

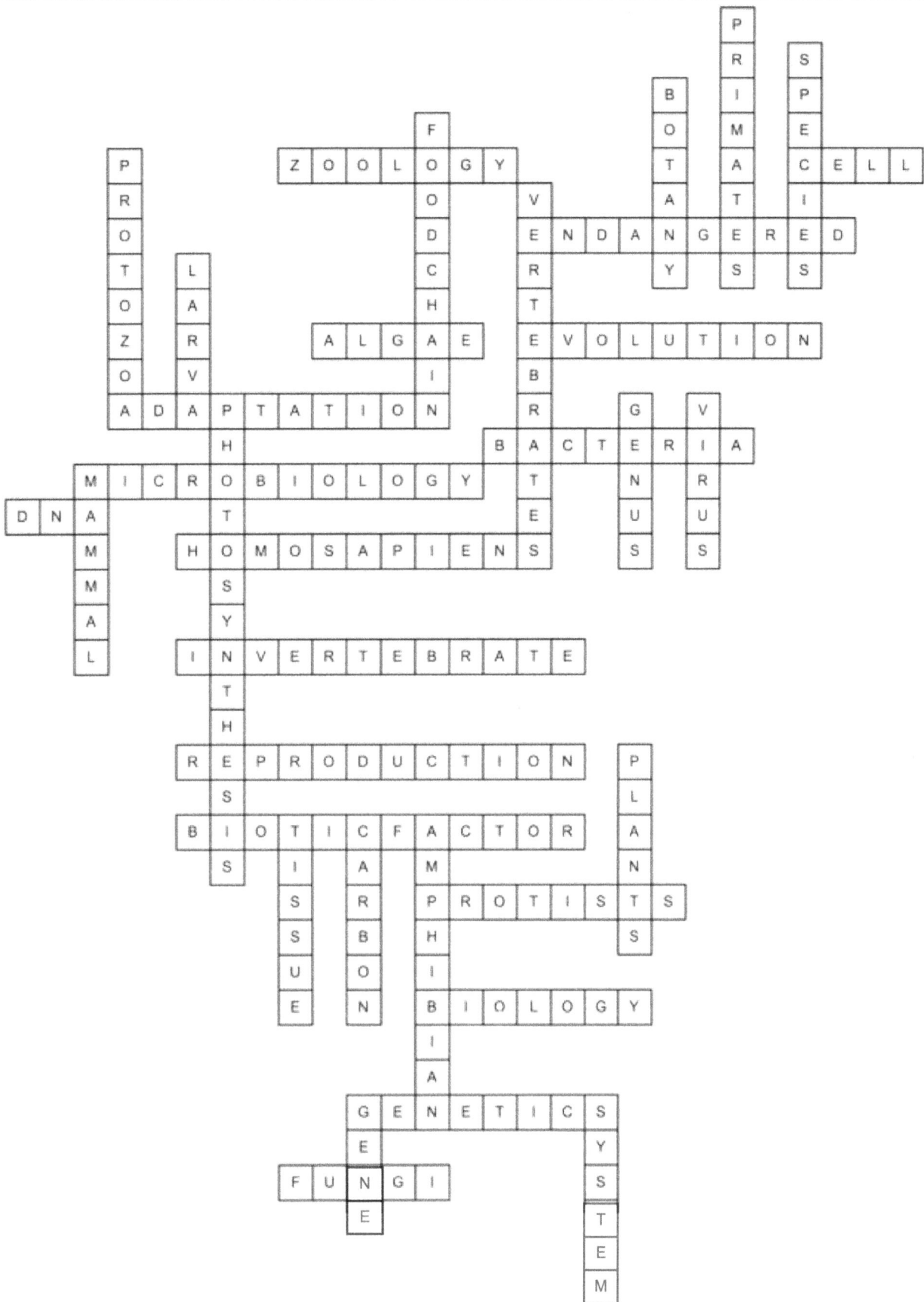

Life Science

A completed crossword puzzle containing the following answers:

Across: ZOOLOGY, CELL, ENDANGERED, ALGAE, EVOLUTION, ADAPTATION, BACTERIA, MICROBIOLOGY, DNA, HOMOSAPIENS, INVERTEBRATE, REPRODUCTION, BIOTICFACTOR, PROTISTS, BIOLOGY, GENETICS, FUNGI

Down: PRIMATES, SPECIE, BOTANY, PROTOZOO, LARVA, FOODCHAIN, VERTEBRAE, GENUS, VIRUS, MAMMAL, PHOTOSYNTHESIS, TISSUE, CARBON, MORPHISM, PLANTS, GENE, SYSTEM

Physical Science

Astronomy

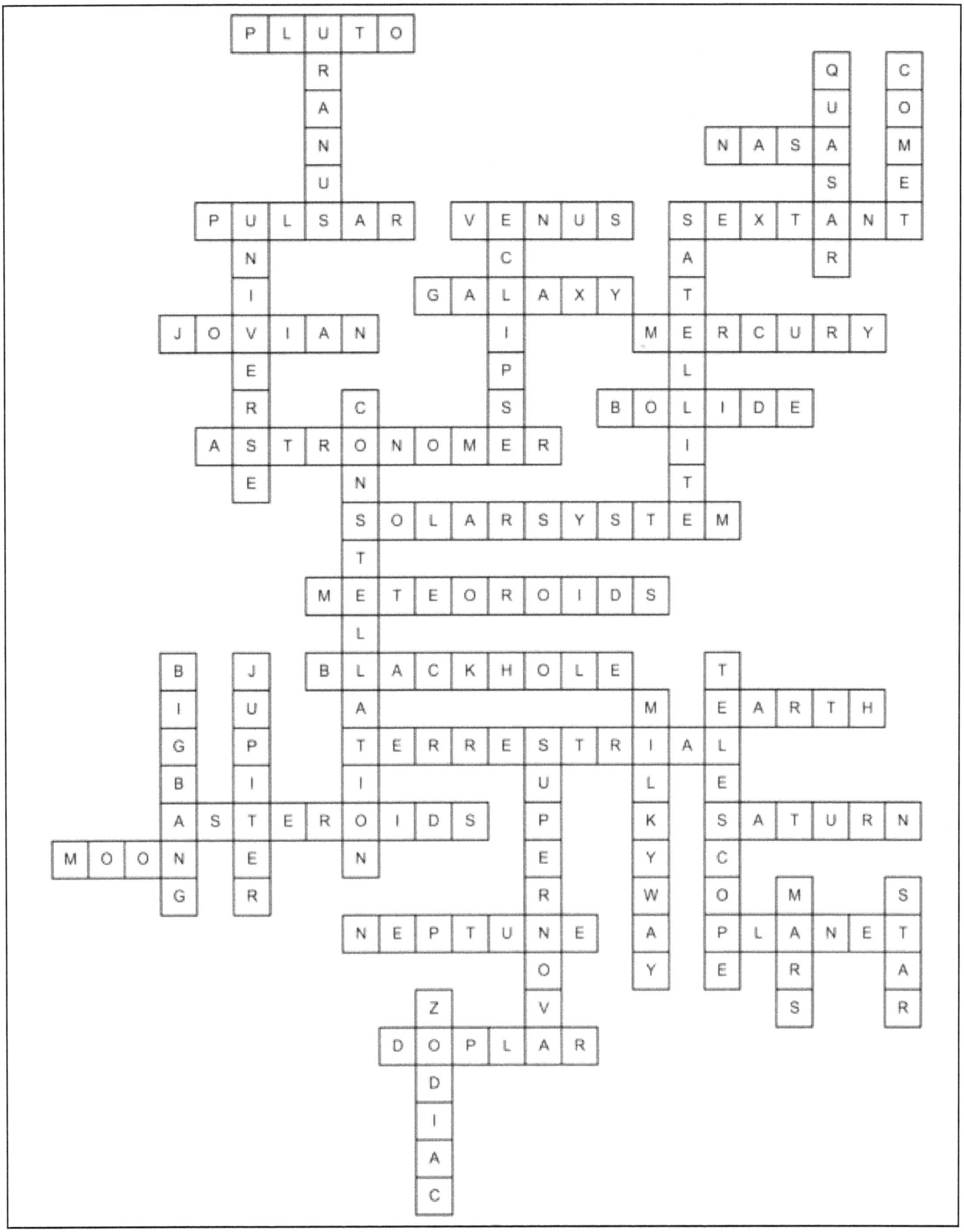
PLUTO
URANUS
QUASAR
COMET
NASA
PULSAR
VENUS
SEXTANT
UNIVERSE
ECLIPSE
SATELLITE
GALAXY
JOVIAN
MERCURY
BOLIDE
ASTRONOMER
CONSTELLATION
SOLARSYSTEM
METEOROIDS
BIGBANG
JUPITER
BLACKHOLE
TELESCOPE
MILKYWAY
EARTH
TERRESTRIAL
SUPERNOVA
ASTEROIDS
SATURN
MOON
MARS
STAR
NEPTUNE
PLANET
ZODIAC
DOPLAR

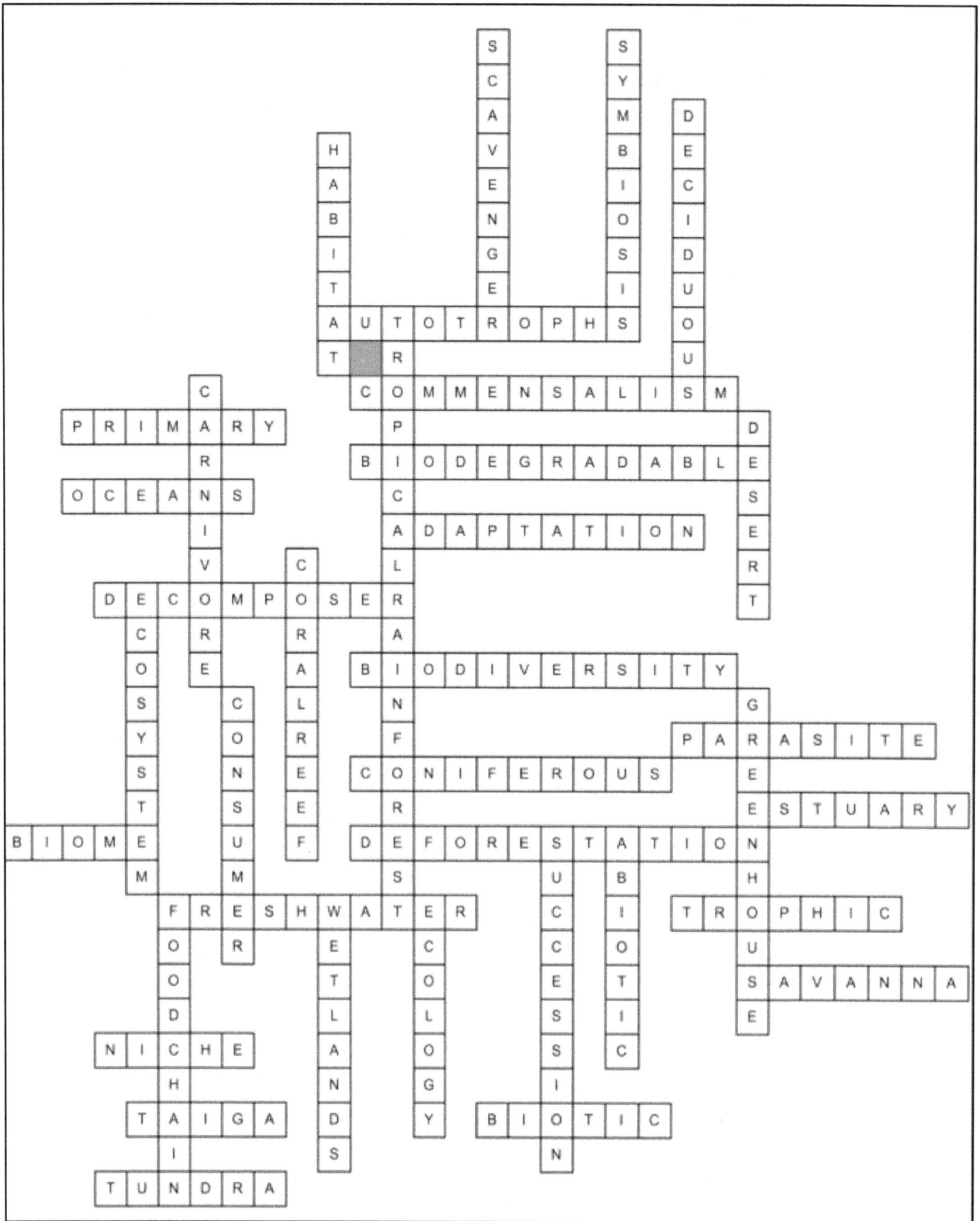

Ecology

A completed crossword puzzle with the following answers:

Across:
- AUTOTROPHS
- COMMENSALISM
- PRIMARY
- BIODEGRADABLE
- OCEANS
- ADAPTATION
- DECOMPOSER
- BIODIVERSITY
- PARASITE
- CONIFEROUS
- ESTUARY
- BIOME
- DEFORESTATION
- FRESHWATER
- TROPHIC
- SAVANNA
- NICHE
- TAIGA
- BIOTIC
- TUNDRA

Down:
- SCAVENGE
- SYMBIOSIS
- DECIDUOUS
- HABITAT
- CARNIVORE
- CONSUMER
- CORAL REEF
- ECOSYSTEM
- COMMUNITY
- FOOD
- WETLANDS
- ECOLOGY
- SUCCESSION
- ABIOTIC
- DESERT
- GENE
- TROPHIC HOUSE

Oceanography

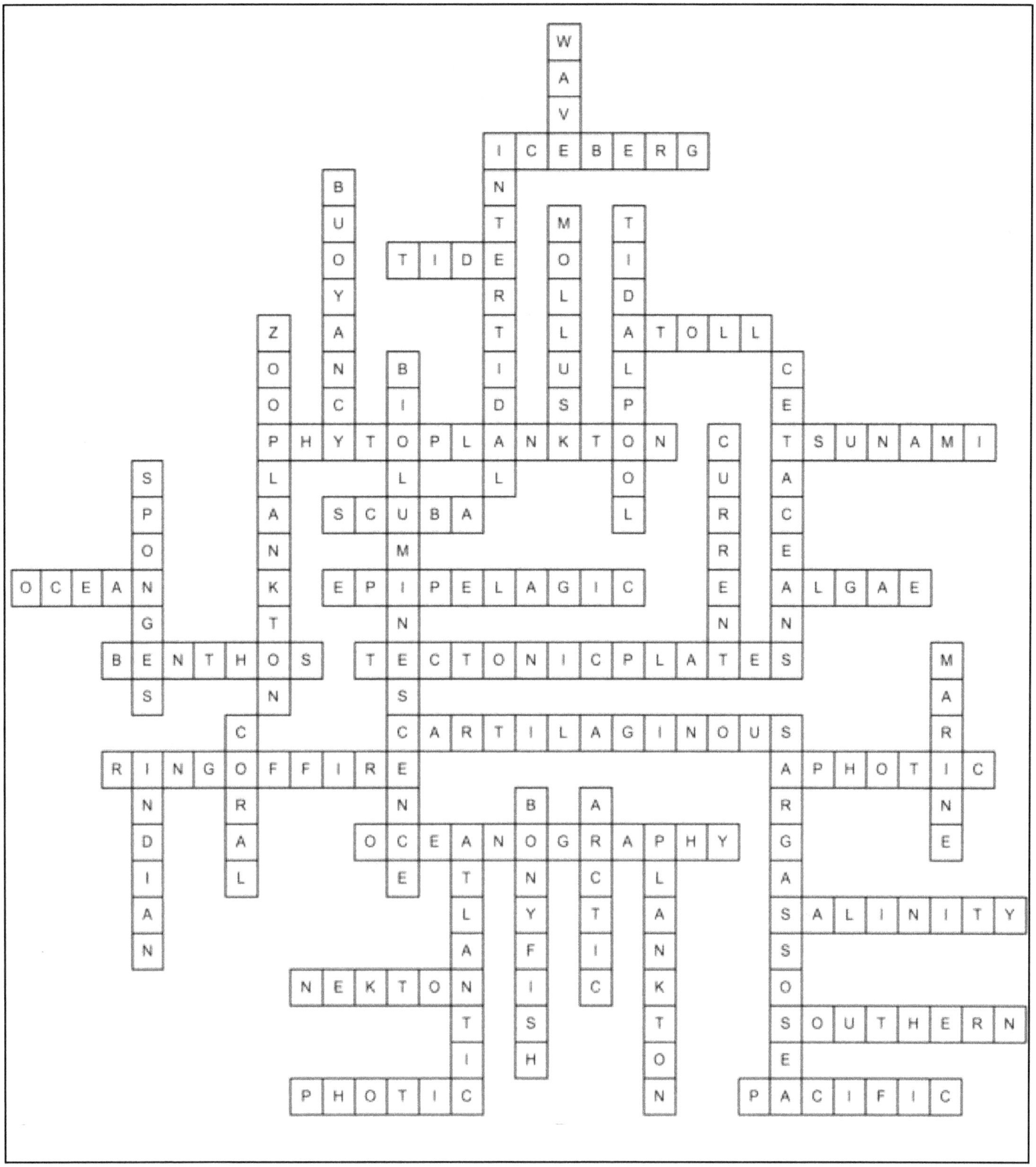
ICEBERG
TIDE
ATOLL
PHYTOPLANKTON
TSUNAMI
SCUBA
OCEAN
EPIPELAGIC
ALGAE
BENTHOS
TECTONICPLATES
CARTILAGINOUS
RINGOFFIRE
APHOTIC
OCEANOGRAPHY
SALINITY
NEKTON
SOUTHERN
PHOTIC
PACIFIC
WAVE
INTERTIDAL
BUOYANCY
MOLLUSK
TIDALPOOL
ZOOPLANKTON
BIOLUMINESCENCE
CETACEANS
CURRENT
SPONGES
MARINE
CORAL
INDIAN
SARGASSOSEA
ATLANTIC
BONYFISH
ARCTIC
PLANKTON

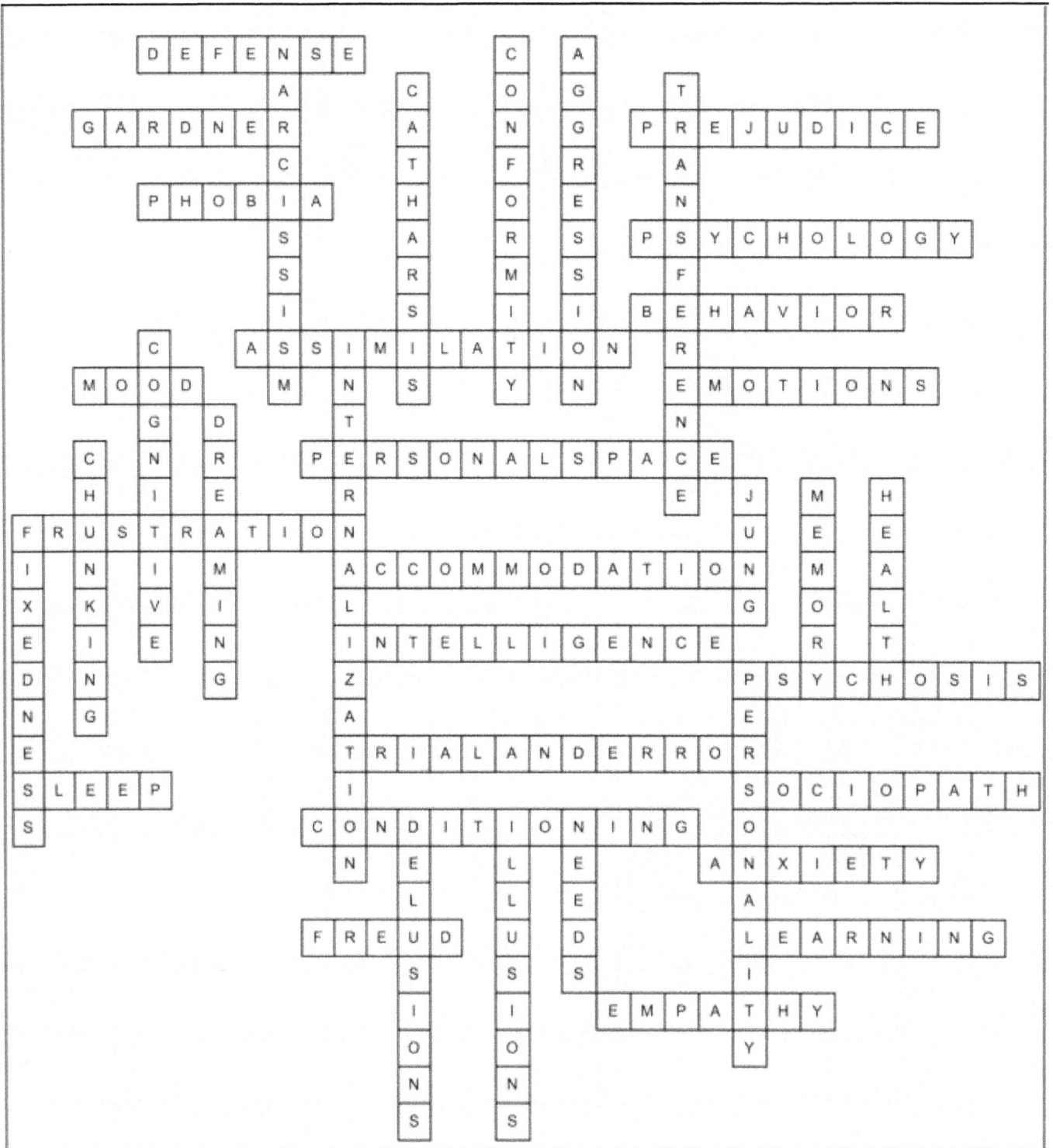

Psychology

A completed crossword puzzle with the following filled-in answers:

- DEFENSE
- GARDNER
- PHOBIA
- PREJUDICE
- PSYCHOLOGY
- BEHAVIOR
- ASSIMILATION
- MOOD
- EMOTIONS
- PERSONALSPACE
- FRUSTRATION
- ACCOMMODATION
- INTELLIGENCE
- PSYCHOSIS
- TRIALANDERROR
- SLEEP
- SOCIOPATH
- CONDITIONING
- ANXIETY
- FREUD
- LEARNING
- EMPATHY

Down answers visible:
- NARCISSISM
- CATHARSIS
- CONFORMISI
- AGGRESSION
- TRANSFERENCE
- COGNITIVE
- DREAMING
- CHING
- FIXEDNESS
- CENTRALIZATION
- JUNG
- MEMORY
- HEALTH
- PEERSONALITY
- CONVULSIONS
- DELUSIONS
- FREUD

Genetics and Heredity

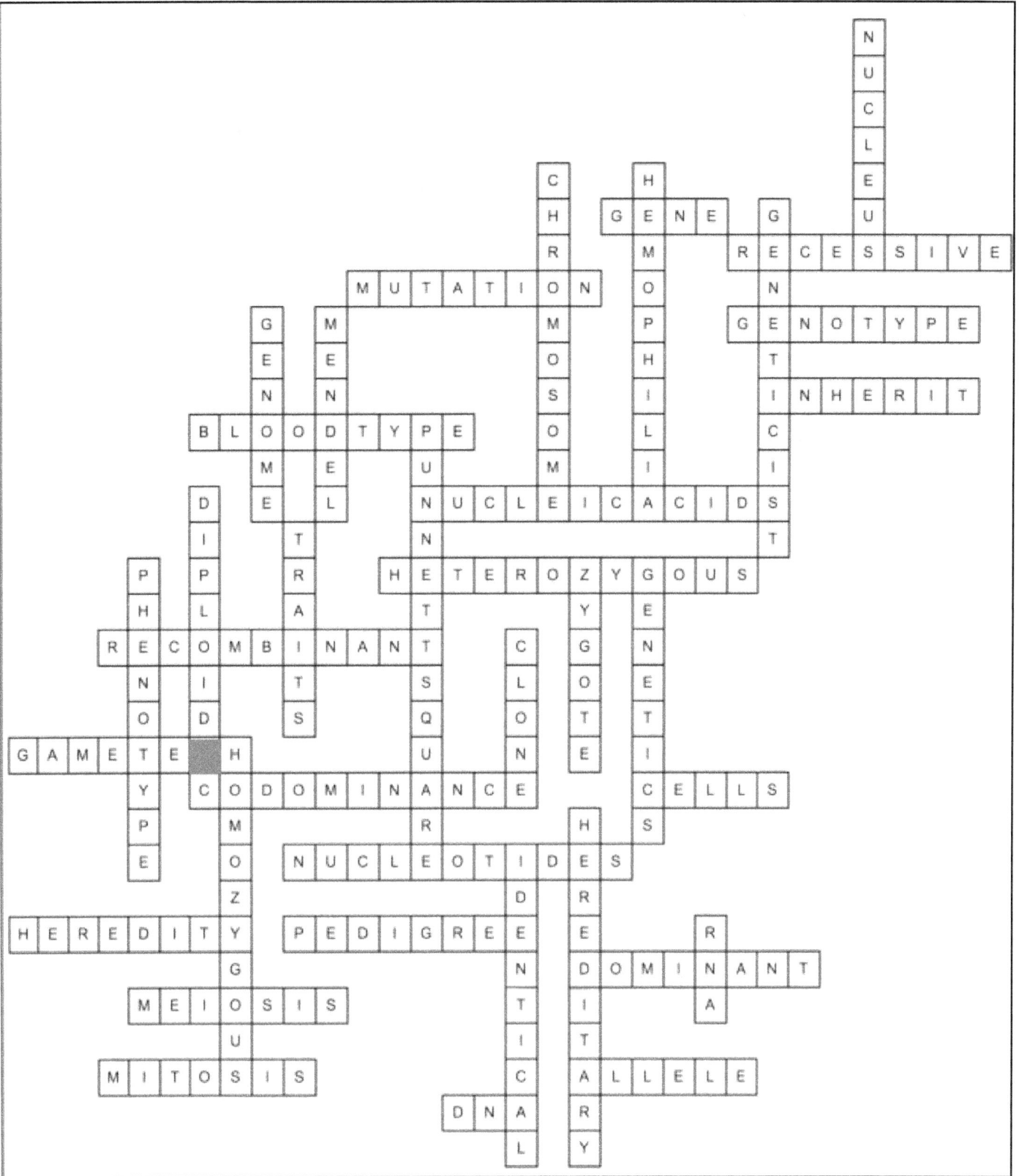

A completed crossword puzzle with the following answers:

NUCLEUS (down)
GENE
GMENT (down - GMENT/GENT)
RECESSIVE
MUTATION
CHROMOSOM (down - CHROMOSOME)
HOMOPHILI (down - HEMOPHILIA)
GENOTYPE
INHERIT
BLOODTYPE
GENE (down)
MENDEL (down)
NUCLEIC ACIDS
DIPLOID (down)
HETEROZYGOUS
PHENOTYPE (down)
RECOMBINANT
GAMETE
CODOMINANCE
CLONE (down)
ZYGOTE (down)
GENETIC (down)
CELLS
NUCLEOTIDES
HEREDITY
PEDIGREE
DOMINANT
MEIOSIS
RECESSIVE TRAIT (down - RECESSIVE TRAIT/DENTRICT... HEREDITARY)
MITOSIS
ALLELE
DNA

Crime Scene Investigation

The puzzle grid contains the following answers:

AFIS

RIGORMORTIS
FINGERPRINT
LIVERMORTIS
ANTEMORTEM

IMPRESSION
FISH
COLD
ALGORMORTIS
DECOMPOSE
AUTOPSY
FORENSICS
TRAJECTORY

BALLISTICS
CALIBER
GUNSHOTRESIDUE
MEDS

CORPUSDELICTI
CHAINOFCUSTODY

SUSPECT
BLOODSPLATTER
LIGATURE

EXCHANGE
CONTAMINATION
TRACE

CRIMINALISTICS

CHROMATOGRAPHY
CODIS
DNA
EVIDENCE
CORONER
LATENT
LUMINAL
FORGERY
ENTOMOLOGY

The Human Body

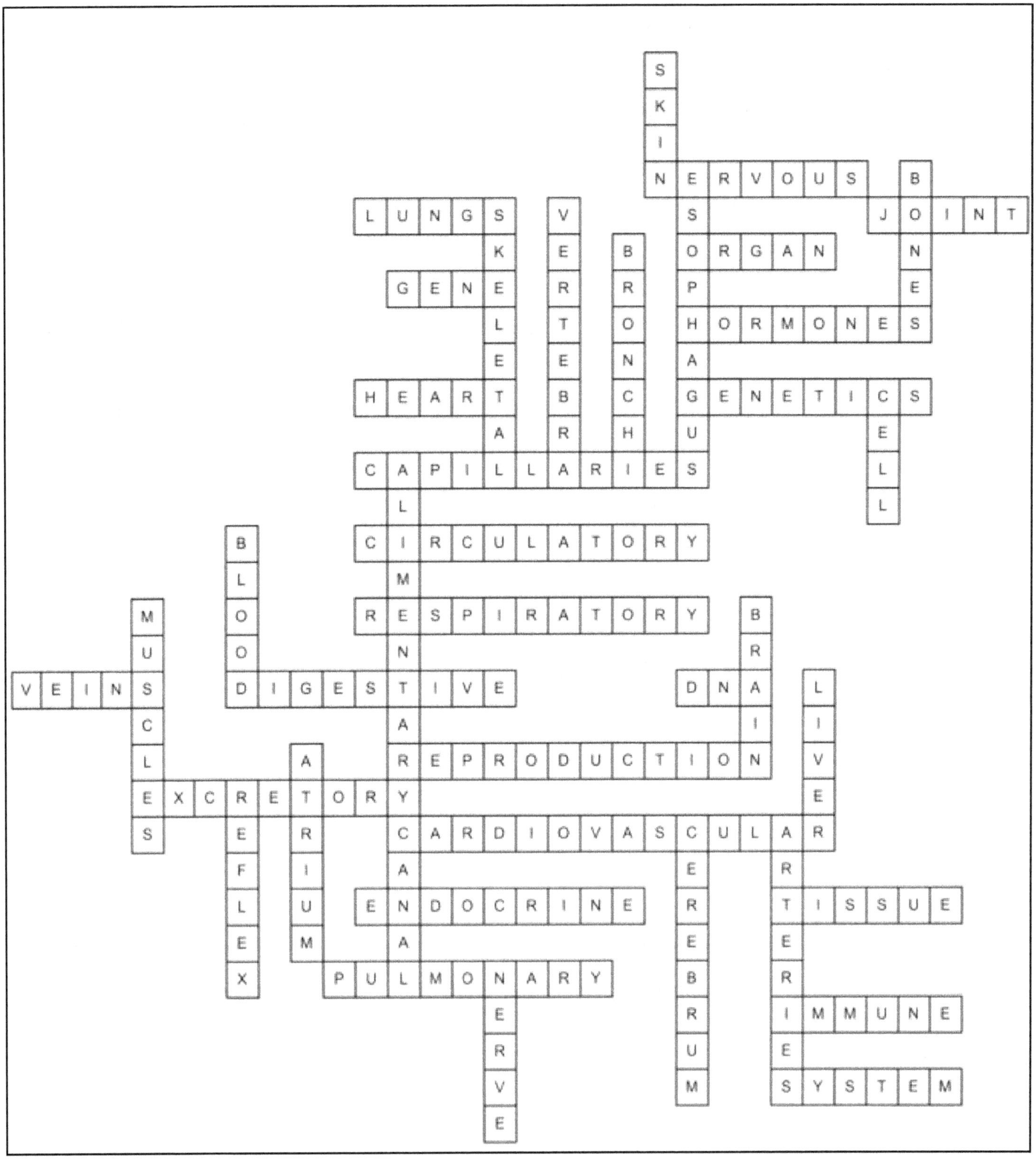
NERVOUS
LUNGS
JOINT
ORGAN
GENE
HORMONES
HEART
GENETICS
CAPILLARIES
CIRCULATORY
RESPIRATORY
VEINS
DIGESTIVE
DNA
REPRODUCTION
EXCRETORY
CARDIOVASCULAR
ENDOCRINE
TISSUE
PULMONARY
IMMUNE
SYSTEM
SKIN
ESOPHAGUS
BONES
SKELETAL
VERTEBRA
BRONCHI
CELL
ALIMENTARY CANAL
BLOOD
MUSCLES
BRAIN
LIVER
ATRIUM
REFLEX
NERVE
CEREBRUM
ARTERIES

Ecosystems Word Search

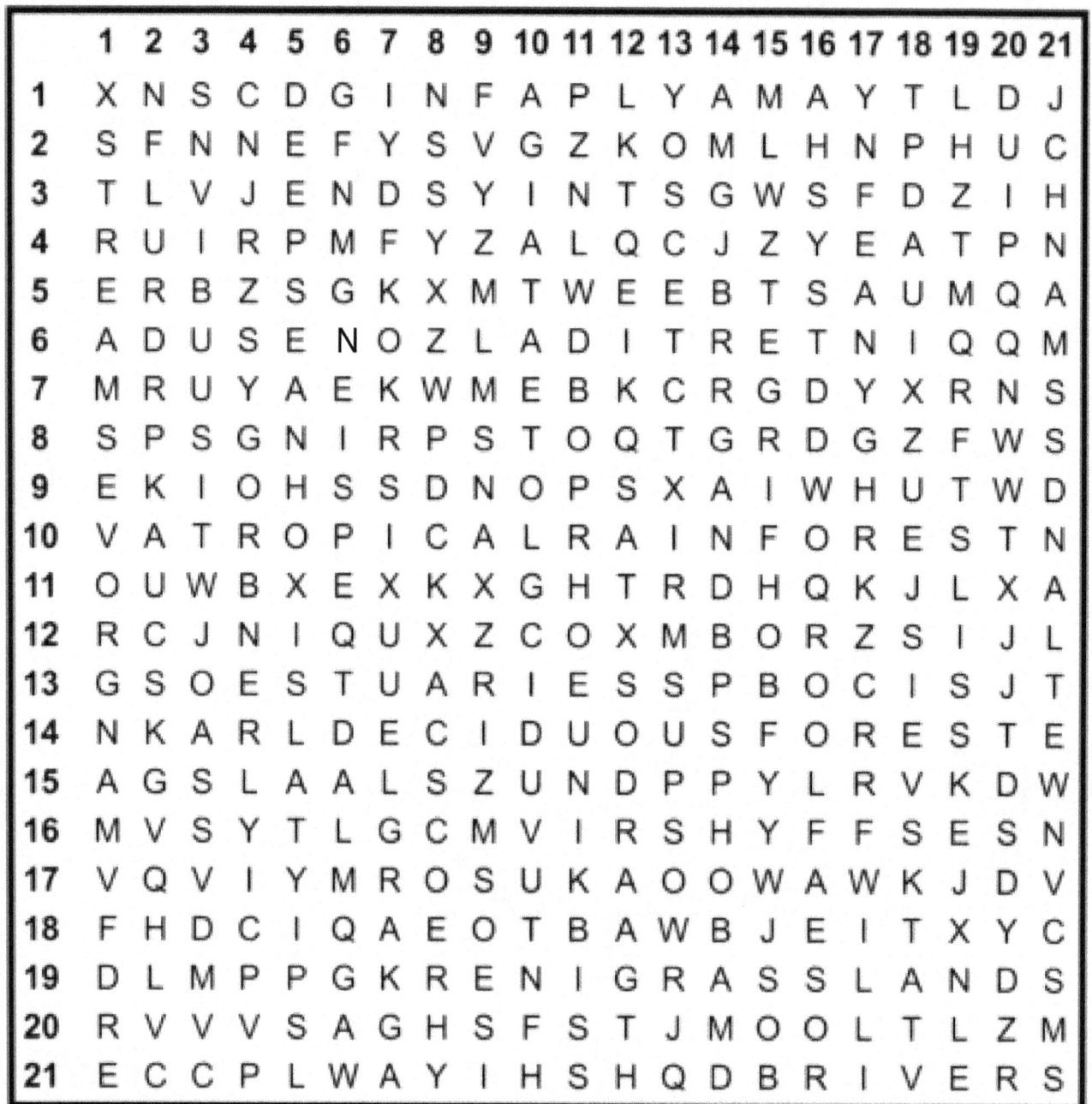

	1	2	3	4	5	6	7	8	9	10	11	12	13	14	15	16	17	18	19	20	21
1	X	N	S	C	D	G	I	N	F	A	P	L	Y	A	M	A	Y	T	L	D	J
2	S	F	N	N	E	F	Y	S	V	G	Z	K	O	M	L	H	N	P	H	U	C
3	T	L	V	J	E	N	D	S	Y	I	N	T	S	G	W	S	F	D	Z	I	H
4	R	U	I	R	P	M	F	Y	Z	A	L	Q	C	J	Z	Y	E	A	T	P	N
5	E	R	B	Z	S	G	K	X	M	T	W	E	E	B	T	S	A	U	M	Q	A
6	A	D	U	S	E	N	O	Z	L	A	D	I	T	R	E	T	N	I	Q	Q	M
7	M	R	U	Y	A	E	K	W	M	E	B	K	C	R	G	D	Y	X	R	N	S
8	S	P	S	G	N	I	R	P	S	T	O	Q	T	G	R	D	G	Z	F	W	S
9	E	K	I	O	H	S	S	D	N	O	P	S	X	A	I	W	H	U	T	W	D
10	V	A	T	R	O	P	I	C	A	L	R	A	I	N	F	O	R	E	S	T	N
11	O	U	W	B	X	E	X	K	X	G	H	T	R	D	H	Q	K	J	L	X	A
12	R	C	J	N	I	Q	U	X	Z	C	O	X	M	B	O	R	Z	S	I	J	L
13	G	S	O	E	S	T	U	A	R	I	E	S	S	P	B	O	C	I	S	J	T
14	N	K	A	R	L	D	E	C	I	D	U	O	U	S	F	O	R	E	S	T	E
15	A	G	S	L	A	A	L	S	Z	U	N	D	P	P	Y	L	R	V	K	D	W
16	M	V	S	Y	T	L	G	C	M	V	I	R	S	H	Y	F	F	S	E	S	N
17	V	Q	V	I	Y	M	R	O	S	U	K	A	O	O	W	A	W	K	J	D	V
18	F	H	D	C	I	Q	A	E	O	T	B	A	W	B	J	E	I	T	X	Y	C
19	D	L	M	P	P	G	K	R	E	N	I	G	R	A	S	S	L	A	N	D	S
20	R	V	V	V	S	A	G	H	S	F	S	T	J	M	O	O	L	T	L	Z	M
21	E	C	C	P	L	W	A	Y	I	H	S	H	Q	D	B	R	I	V	E	R	S

The terms below are listed with their starting row and column.

CORAL REEFS 12:2

DECIDUOUS FOREST 14:6

DEEP SEA 1:5

DESERTS 3:18

ESTUARIES 13:4

GRASSLANDS 19:12

INTERTIDAL ZONES 6:18

LAGOONS 14:5

LAKES 21:5

MANGROVES 16:1

PONDS 9:11

RIVERS 21:16

SALT MARSH 13:2

SEA FLOOR 19:16

SPRINGS 8:9

STREAMS 2:1

TAIGA 5:10

TROPICAL RAIN FOREST 10:3

TUNDRA 4:19

WETLANDS 15:21

Science Crossword Puzzles: Grades 6 & Up

Hidden Body Parts

Find the hidden body part in each sentence. A clue is provided for each.

1. No**el bow**ed to the king.
Clue: Joint between the forearm and the upper arm

2. Jac**k nee**ds a bandage for his wound.
Clue: Joint between the thigh and the lower leg.

3. **Hank le**d the team in home runs.
Clue: Joint connecting the foot with the leg

4. Sam asked his mother, "**Should Er**ic or I go to the store?"
Clue: Upper joint of the human arm

5. Because of the strong **wind, Pipe**r did not open her umbrella.
Clue: What the trachea is sometimes called

6. The child was fascinated by the **colon**y of ants.
Clue: What the large intestine is called

7. The **liver**y car arrived late, so the couple almost missed their flight.
Clue: Organ that secretes bile and stores fat and sugar

8. Emma watched intently as two b**ears** frolicked in her backyard.
Clue: Organs that perceive sound by detecting vibrations

9. Sandy does not like sweet or sour food.
Clue: Body without head, neck, and limbs

10. Sir **Is**aac Newton is known for his laws of motion and universal gravitation.
Clue: Colored, ring-shaped membrane behind the cornea of the eye

11. When told to write about an adventu**re, Tina** wrote about rock climbing.
Clue: Layer at the back of the eyeball containing cells that are sensitive to light

12. "Chri**s, kin**dly shut the door on your way out," the shopkeeper requested.
Clue: Organ forming the natural outer covering of the body

Volcanic Eruptions Puzzle

Complete the puzzle. Then write the letters written on the larger emboldened lines on the lines at the left. The result will be a word describing "rocks formed from the solid fragments ejected during a volcanic eruption."

__P__   A very light and porous volcanic rock   **P** U M I C E

__Y__   A type of hot spring   G E **Y** S E R

__R__   A large cave   C A V E **R** N

__O__   A lump of lava thrown out by a volcano   B **O** M B

__C__   Earth's is mostly iron   **C** O R E

__L__   Rock fragments ejected from volcano   **L** A P I L L I

__A__   Powdery material thrown out by a volcano   **A** S H

__S__   Hard, dark, glasslike volcanic rock   O B **S** I D I A N

__T__   A volcano is one in the Earth's surface   V E N **T**

__I__   Rocks formed by cooling & solidifying of magma or lava   **I** G N E O U S

__C__   Combustible rock; consists mainly of carbonized plant matte   **C** O A L

Optional Lists of Words and Terms

These lists are provided for your convenience. If a puzzle is used as an introduction or just for fun, you might want to provide the list of words. On the other hand, if the puzzle is being done in lieu of a quiz, you might choose not to utilize them. In either case, solutions to the puzzles are provided.

Earth Science

aquifer   atmosphere   bedrock   caldera   canyon   cavern   continental drift   core
crust   earthquake   erosion   fault   folds   fumaroles   geology   geyser   glacier
hydrologic cycle   hydrosphere   lithosphere   mantle   minerals   molten rock   mountains
plateaus   Richter   Ring of Fire   rocks   seismic waves   tectonic plates   tephra
terrestrial   tides   troposphere   tsunami   volcano   weathering

---

Life Science

adaptation   algae   amphibian   bacteria   biology   biotic factor
botany   carbon   fungi   gene   genetics   genus   *homo sapiens*
invertebrate   larva   mammal   microbiology   photosynthesis   plants
primates   protists   protozoa   reproduction   species
system   tissue   vertebrates   virus   zoology

Physical Science

acoustics   aerodynamics   atom   buoyancy   chemistry   color   conductor
convection   density   electricity   energy   force   frequency   friction
fusion   gear   generator   insulator   joule   kinetic   light   magnet   mass
matter   molecule   nuclear   nucleus   optics   physics   potential   power
prism   reflection   refraction   spectrum   temperature   thermal   watt   waves

---

Astronomy

asteroids   astronomer   big bang   black hole   bolide   comet   constellation
Doplar   Earth   eclipse   galaxy   Jovian   Jupiter   meteoroids   Mars
Mercury   Milky Way   moon   NASA   Neptune   planet   Pluto   pulsar
quasar   satellite   Saturn   sextant   solar system   star   supernova
telescope   terrestrial   universe   Uranus   Venus   zodiac

Ecology

abiotic   adaptation   autotrophs   biodegradable   biodiversity   biome   biotic
carnivore   coniferous   commensalism   consumer   coral reef   deciduous   decomposer
deforestation   desert   ecology   ecosystem   estuary   food chain   freshwater   greenhouse
habitat   niche   oceans   parasite   primary producers   savanna   scavenger
succession   symbiosis   taiga   trophic   tropical rainforest   tundra   wetlands

---

Oceanography

algae   aphotic   Arctic   Atlantic   atoll   benthos   bioluminescence   bony fish
buoyancy   cartilaginous   cetaceans   coral   current   epipelagic   iceberg   Indian
intertidal   marine   mollusk   nekton   ocean   oceanography   Pacific   photic
phytoplankton   plankton   Ring of Fire   salinity   Sargasso Sea   scuba   Southern
sponges   tectonic plates   tidal pool   tide   tsunami   wave   zooplankton

Psychology

accommodation   aggression   anxiety   assimilation   behavior   catharsis   chunking
conditioning   cognitive   conformity   defense   delusions   dreaming   emotions   empathy
fixedness   Freud   frustration   Gardner   health   illusions   intelligence   internalization
Jung   learning   memory   mood   narcissism   needs   personality   personal space
phobia   prejudice   psychology   psychosis   sleep   sociopath   trial and error   transference

Genetics and Heredity

allele   blood type   cells   chromosome   clone   codominance   diploid   DNA
dominant   gamete   gene   geneticist   genetics   genome   genotype   hemophilia
hereditary   heredity   heterozygous   homozygous   identica   inheri   meiosis
Mendel   mitosis   mutation   nucleic acids   nucleotides   nucleus   pedigree
phenotype   Punnett square   recombinant   recessive   RNA   traits   zygote

Crime Scene Investigation

AFIS   *algor mortis*   antemortem   autopsy   ballistics   barrier tape   blood splatter   caliber
chain of custody   chromatography   CODIS   cold   contamination   coroner   *corpus delicti*
criminalistics   criminology   DNA   entomology   evidence   exchange   fingerprints   FISH   forensics
forgery   gunshot residue   impression   latent   ligature   *liver mortis*   luminal   manner of death
medical examiner   *modus operandi*   odontology   perpetrator   *rigor mortis*   suspect   trace   trajectory

The Human Body

alimentary canal   arteries   atrium   blood   bones   brain   bronchi   capillaries
cardiovascular   cel   cerebrum   circulatory   DNA   digestive   endocrine
esophagus   excretory   gene   genetics   heart   hormones   immune   liver   lungs
joint   muscles   nerve   nervous   organ   pulmonary   respiratory
reflex   reproduction   skeletal   skin   system   tissue   veins   vertebra